Liquid Electrolyte Chemistry for Lithium Metal Batteries

Liquid Electrolyte Chemistry for Lithium Metal Batteries

Design, Mechanisms, Strategies

Jianmin Ma

WILEY-VCH

Author

Prof. Jianmin Ma
University of Electronic Science and
Technology of China
School of Materials and Energy
No. 2006, Xiyuan Ave
West Hi-Tech Zone
611731 Chengdu
China

Cover Image: © AiMstock/Shutterstock

Library of Congress Card No.: applied for

British Library Cataloguing-in-Publication Data
A catalogue record for this book is available from the British Library.

Bibliographic information published by the Deutsche Nationalbibliothek
The Deutsche Nationalbibliothek lists this publication in the Deutsche Nationalbibliografie; detailed bibliographic data are available on the Internet at <http://dnb.d-nb.de>.

Print ISBN: 978-3-527-35014-8
ePDF ISBN: 978-3-527-83637-6
ePub ISBN: 978-3-527-83638-3
oBook ISBN: 978-3-527-83639-0

Typesetting Straive, Chennai, India
Printing CPI Group (UK) Ltd, Croydon CR0 4YY

Printed on acid-free paper

C118017_310122

Contents

Preface

At present, we have been facing pressing environmental challenges due to temperature rise. This rise gives us an urgent task to decrease CO_2 emission through technology innovations. Especially, oil-fueled cars should be replaced by electric vehicles, which require high-energy-density rechargeable batteries. In recent years, lithium metal batteries are undergoing redevelopment due to the emergence of novel technologies, which may solve risks originating from lithium anodes in lithium metal batteries. One of the important technologies is based on the development of electrolytes used in lithium metal batteries. With this background, we have written in this book about the concepts, development, and challenges of electrolytes in lithium metal batteries.

Electrolyte is an important issue for any energy storage device; however, it is important for lithium metal batteries from the viewpoint of task issue. In addition, the cycling performance and coulombic efficiency of lithium metal batteries are involved. Therefore, some issues at present will be studied, i.e. solid electrolyte interphase, cathode electrolyte interphase, high-voltage electrolytes, safe electrolytes, high-temperature electrolytes, low-temperature electrolytes, and high-concentration electrolytes. Thus, this book discusses these topics from the types of lithium metal batteries, electrode–electrolyte interphases, high-voltage electrolytes, extreme temperature electrolytes, safe electrolytes, and high-concentration electrolytes. I believe that these contents will enrich the development of electrolytes for lithium metal batteries.

Chapter	Main contributors
Chapter 1 on lithium metal batteries	Dr. Fang Li, Prof. Jianmin Ma
Chapter 2 on electrode–electrolyte interphase	Dr. Mingguang Wu, Prof. Jianmin Ma
Chapter 3 on safe electrolytes	Dr. Huaping Wang, Prof. Jianmin Ma
Chapter 4 on high voltage electrolytes	Dr. Shihan Qi, Prof. Jianmin Ma
Chapter 5 on extreme temperature electrolytes	Dr. Junda Huang, Prof. Jianmin Ma
Chapter 6 on high concentration electrolytes	Dr. Gaoxue Jiang, Prof. Jianmin Ma
Chapter 7 on lithium metal batteries	Dr. Jiandong Liu, Mr. Jian He, Mr. Chunlei Zhu, Prof. Jianmin Ma
Chapter 8 on outlook	Dr. Daxiong Wu, Mr. Jian He, Prof. Jianmin Ma

We would also like to extend our thanks to Mr. Kanglong Guo, Mr. Zhongsheng Wang, Dr. Yulu Yang, Dr. Xi Tang, Mr. Nan Jiang, Mr. Zhengxuan Yang, and Dr. Chuan Wang for their comments and revisions for the entire book. Though this book was completed in a very short time, utmost care has been taken to ensure accuracy; however, if there are any errors in this book, I hope that the readers, students from all grades to researchers, will understand us. Finally, we hope that this book gives the readers some useful knowledge about electrolytes and help them to familiarize with and study the electrolytes in this field.

15 August 2021

Jianmin Ma
University of Electronic Science and Technology of China,
School of Materials and Energy
Chengdu, China

1

Lithium Metal Batteries

1.1 History

Li metal has been widely regarded as the ideal anode material due to its ultrahigh theoretical specific capacity (3860 mAh g^{-1}) and very low redox potential (~3.040 V vs. standard hydrogen electrode) [1, 2]. The first visible Li metal secondary battery was operated by Stanley Whittingham at Exxon in the 1970s. In the late 1980s, Li metal batteries based on MoS_2 cathode and excess Li were commercialized and sold to the market, which could be cycled hundreds of times. But due to safety accidents, including fires caused by Li dendrites, all the cells were recalled [3]. Since then, safety concerns about Li metal anodes still prevail. Although NEC and Mitsui realized an ultra-long cycling stability of Li metal cells over 500 000 cycles, the safety issue is not solved. Later, Sony Company developed a safe Li-ion battery based on a carbonaceous anode instead of Li metal, which can overcome the safety concerns related to Li dendrites, and these kinds of cells were commercialized and are in use until today [4]. As a result, the studies on Li metal anodes were halted.

Now, with the emergence of vehicle devices and hybrid grids, the state-of-the-art Li ion batteries based on graphite anodes are reaching their theoretical capacity, which cannot fulfill the increasing demand for high energy density [5]. Therefore, the application of Li metal as anode attracts the researchers. Over the four decades of research on Li metal anodes, great achievements have been made in improving the energy density and cycling stability of Li metal batteries [6]. Among them, various cathode alternatives have been reported, such as intercalation cathodes [7], sulfur/selenium/tellurium composites [8–10], oxygen/air electrodes [11], iodine/bromine flow cathode [12], and lithium-free transition metal oxides [13]. These batteries have their own advantages, but also exhibit their intrinsic problems, which hamper practical applications of Li metal batteries. But these battery systems all have common problems related to Li metal [14]. First, Li metal can react with the electrolyte to form a solid electrolyte interphase (SEI) layer on the surface, which will consume the electrolyte and increase the internal resistance. Second, Li dendrites will grow and cause safety concerns. Third, large volume change of Li metal can break the SEI and lead to further reaction between Li metal and the electrolyte, resulting in the failure of cells. All these disadvantages need to be addressed before the commercialization of Li metal batteries.

Liquid Electrolyte Chemistry for Lithium Metal Batteries: Design, Mechanisms, Strategies,
First Edition. Jianmin Ma.

1.2 Types

In this chapter, we summarize several different types of Li metal batteries, such as lithium–oxygen batteries, lithium–sulfur batteries, lithium–selenium/tellurium batteries, lithium–iodine/bromine batteries, and lithium-free transition metal oxide Li batteries. In each section, we first introduce the working mechanisms, then the design strategies for cathode structures, the modification of electrolyte, and the protection methods for Li metal anodes.

1.2.1 Lithium–Oxygen Batteries

With the continuous consumption of nonrenewable fossil fuels and new requirements for carbon neutrality, green energy systems have been paid much attention. Among them, Li–O_2 batteries using O_2 as cathode and Li metal as anode have the highest energy density (~3500 Wh kg^{-1}). The cathode, O_2, can be inexhaustibly extracted from open air and show the lightweight property to reduce the whole weight of Li–O_2 cells. Unlike cathodes in Li–ion batteries, that consist of high-cost nickel or/and cobalt metal, the abundance and cheap O_2 cathodes can dramatically reduce the cost of Li–O_2 batteries. Moreover, the Li anode exhibits high specific capacity and lowest electrochemical potential. Although Li–O_2 batteries have so many advantages over Li–ion batteries, their practical applications are hampered by the sluggish oxygen reduction reactions (ORR) and oxygen evolution reactions (OER) and random deposition of discharge products (Li_2O_2), leading to high overpotentials, low capacity, and unstable cycling performance [15]. In this section, the working mechanism of Li–O_2 batteries is discussed first and then the reasonable design for cathode and anode protection is summarized. Finally, problems are summarized and future directions are provided.

1.2.1.1 Working Mechanism of Li–O_2 Batteries

As shown in Figure 1.1a, a Li–O_2 battery consists of three parts: Li metal anode, electrolyte, and porous cathode [16]. Unlike other energy storage systems, the Li–O_2 battery is a semi-open system, which can allow O_2 penetrate into the cathode side to participate in electrochemical reactions. During discharge, the Li anode will be oxidized to generate Li^+, and the oxygen is reduced on the cathode side, which can react with Li ions to form Li_2O_2, as described by Eq. (1.1):

$$O_2 + 2Li^+ + 2e^- \rightarrow Li_2O_2 \tag{1.1}$$

It looks very simple according to the reaction equation, but this direct reaction seems unlikely to occur electrochemically because the two-electron process requires much higher entropic barriers than that of a one-electron process. Actually, the discharging process is much more complicated than the equation [18], which is shown in Figure 1.1b. O_2 is first adsorbed on the active sites of cathode materials, donated as O_2^*, and then O_2^* will get one electron and reacts with Li^+ to form LiO_2^* (Eq. (1.2)):

$$O_2^* + Li^+ + e^- \rightarrow LiO_2^* \tag{1.2}$$

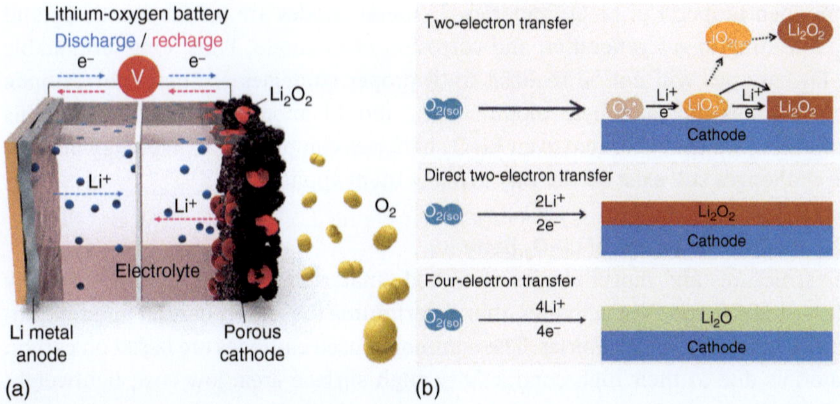

Figure 1.1 (a) Schematic operation proposed for rechargeable aprotic Li–O_2 batteries. Source: Reproduced with permission Kwak et al. [16]. (b) Schematic of the Li_2O_2 growth pathways. Source: Reproduced with permission Lyu et al. [17]. Reproduced with permission of (2017) Royal Society of Chemistry.

After that, the LiO_2^* have three possible reaction pathways to finally transfer to Li_2O_2, as described by the following equation:

$$LiO_2^* + Li^+ + e^- \rightarrow Li_2O_2^* \tag{1.3}$$

$$LiO_2^* + LiO_2^* \rightarrow Li_2O_2^* + O_2^* \tag{1.4}$$

$$LiO_{2(sol)} + LiO_{2(sol)} \rightarrow Li_2O_2 + O_2 \tag{1.5}$$

The LiO_2^* forms $Li_2O_2^*$ through electrochemical reduction (Eq. (1.3)) or disproportionation on the surface (Eq. (1.4)), both of which are referred to as the surface growth model. It has been reported that the electrochemical reduction pathway can be easily realized due to the low kinetic barrier and free energy [19]. The adsorbed LiO_2^* may also diffuse into the solution by modulating the electrolyte, which can be converted into Li_2O_2 via disproportionation in solution (Eq. (1.5)), which is regarded as the solution growth model.

The most expected is the four-electron transfer process, i.e. the O_2 is fully reduced to lithium oxide (Li_2O) because of its higher electron transfer number. However, this reaction pathway is kinetically impossible because Li_2O is not observed by any spectroscopic technique at a cutoff voltage of above 2.0 V. Moreover, during the charging process, the Li_2O could not convert into O_2 [20]. On the cathode side, carbonaceous materials as matrixes can be decomposed on high voltage or react with Li_2O_2 to form Li_2CO_3 [21]. These solid side products will accumulate on the cathode surface, leading to blockage of electron and O_2 channels and the failure of battery. Therefore, cathode materials with suitable pore and relatively high pore volume are desirable. The superoxide as discharge product can react with the electrolyte and form side products on the cathode surface. In addition, due to the intrinsic sluggish ORR/OER process, the rate capability of Li–O_2 batteries is poor and catalysts should be applied on the cathode side [22, 23]. In view of semi-open property of Li–O_2 batteries, Li metal anodes are unstable with O_2 and H_2O, leading to gas generation and corrosion of Li anode. Thus, long-term stable Li–O_2 batteries will not be realized until proper strategies are made for cathode structure design, electrolyte modification, and Li anode protection. From this discussion, we conclude that even Li–O_2 batteries can provide high energy density, the challenges still exist on the way to make them applicable.

1.2.1.2 Cathode Design of Li–O_2 Batteries

The structures and morphologies of cathode materials in Li–O_2 batteries directly affect the ORR and OER activities, thus determining the energy density and coulombic efficiency of Li–O_2 batteries. The commonly used cathodes are based on carbon materials due to their high conductivity, high surface area, low cost, lightweight property, and various and controllable structures [24]. The ideal cathode structure should have high chemical/electrochemical stability in the reversible cycling process; high surface area with mesoscale pores for large Li_2O_2 storage, suitable porosity to control the size of Li_2O_2 product; and high electrical conductivity to reduce the overpotential caused by the insulator and insoluble Li_2O_2 product.

For the cathode design, at the beginning, researchers largely focused on preparing nanosized materials with high conductivity and high surface area. For example, Zhang's group reported a porous air electrode with functionalized graphene sheets (FGSs) with lattice defects and hydroxyl, epoxy, and carboxyl groups (Figure 1.2a) [25]. The prepared FGSs consist of interconnected bimodal pore structures with micro- and nanometer length scales, which facilitate the penetration of O_2 and the transportation of electrons. In addition, the density functional theory (DFT) calculation indicates that the lattice defect sites on the FGSs can help to form small, nanometer-sized discharge products (Li_2O_2). The FGS-based air cathode (with no catalyst) delivers an exceptionally high capacity of 15 000 mAh g^{-1}. In 2016, RuO_2/CNT (carbon nanotube) was synthesized as cathode for Li–O_2 batteries by Zhang's group. Due to the synergistic effect of RuO_2 with good catalysis and CNT with high conductivity and 1D structure, the battery delivered a high discharge capacity of 29 900 mAh g^{-1} and a relatively long cycling stability over 171 cycles. Recently, Sun's group fabricated porous Mo_2C/C nanoflowers assembled by ultrathin carbon nanosheets decorated with Mo_2C quantum dots (Figure 1.2b,c) [26]. As cathode materials, Mo_2C/C materials show excellent catalytic activity with a low discharge/charge gap of 1.2 V in the first cycle (Figure 1.2d). The ultrathin carbon nanosheets could suppress the agglomeration of Mo_2C quantum dots, as well as improve the conductivity of the Mo_2C/C cathode. In addition, the ultrathin carbon nanosheets and porous nanoflowers structure could ensure fast Li ions and O_2 diffusion. The Mo_2C/C delivers a relatively high specific capacity of 7500 mAh g^{-1}. With a cutoff capacity of 500 mAh g^{-1}, the Li–O_2 battery exhibits no obvious capacity fading for 100 cycles (Figure 1.2e). The excellent performance may be ascribed to the synergistic effect of ultrathin carbon nanosheets, highly dispersed Mo_2C quantum dots, and the special porous structure.

But in this period, some binders were used in the preparation of cathodes, which were unstable and could be decomposed during cycling. Therefore, freestanding

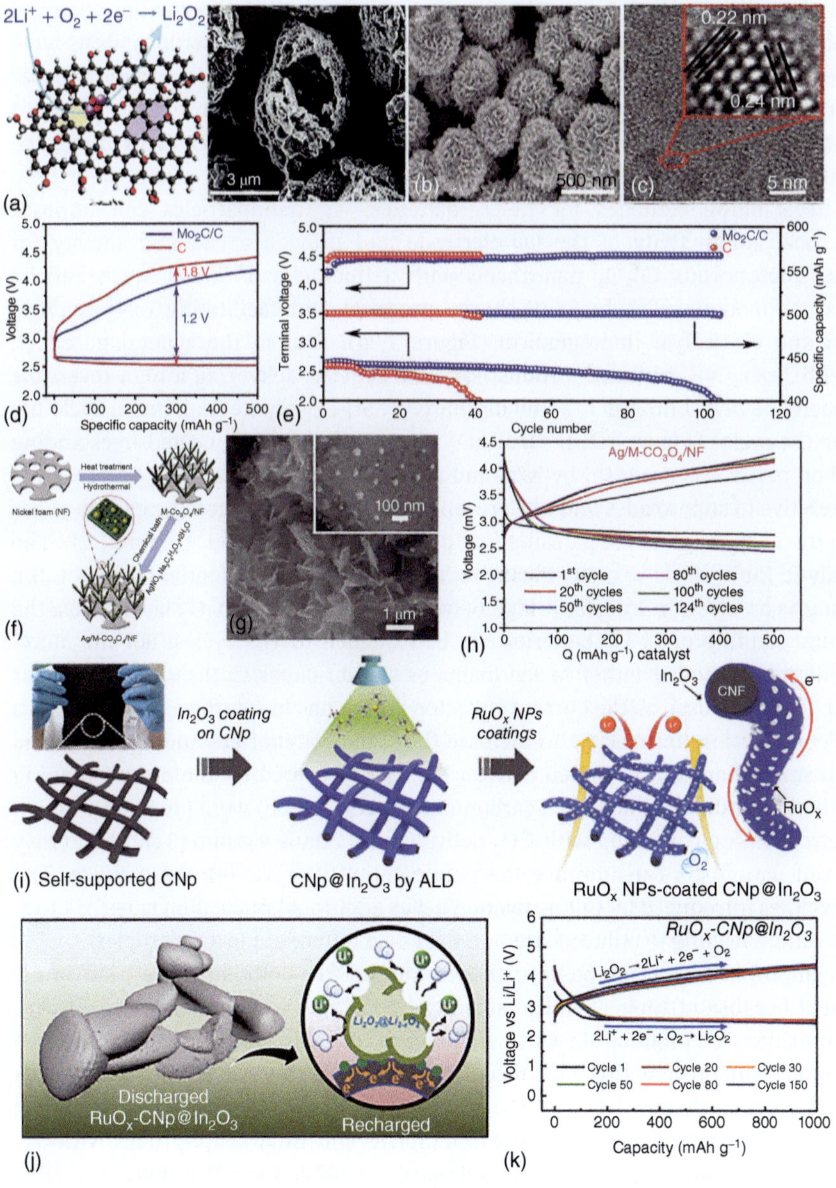

Figure 1.2 (a) Schematic structure of a functionalized graphene sheet with an ideal bimodal porous structure. Source: Xiao et al. [25] / with permission of American Chemical Society (b) SEM and (c) TEM images of Mo_2C/C. (d) The first discharge/charge curves and (e) cycling performance of Mo_2C/C and C as cathodes at 100 mA g^{-1}. Source: Jiao et al. [26] / with permission of Elsevier. (f) Synthetic procedure of Ag/M-Co_3O_4/NF composite structure; (g) SEM image of Ag/M-Co_3O_4/NF composite; (h) discharge/charge curves of Li–O_2 batteries with Ag/M-Co_3O_4/NF electrode at different cycles. Source: Huang et al. [27] / with permission of Elsevier. (i) Schematic illustration of synthesis of RuO_x-CNp@In_2O_3; (j) suggested illustration for Li_2O_2 decomposition in RuO_x-CNp@In_2O_3; (k) Discharge–charge profiles of RuO_x-CNp@In_2O_3 at a current density of 0.2 mA cm^{-2} with a fixed capacity of 1000 mAh g^{-1}. Source: Jung et al. [28] / with permission of Elsevier.

and binder-free cathodes were developed using simple hydrothermal or chemical vapor deposition methods. Various freestanding morphologies with catalysts were reported, such as nanoarrays, nanowalls, nanowires, grown on carbon paper or nickel foam or carbon cloth [29–31]. In 2017, Luo and coworkers prepared Ag-decorated highly mesoporous Co_3O_4 nanosheets on nickel foam using the hydrothermal and chemical bath method (Figure 1.2f) [27], which were used as freestanding cathodes for Li–O_2 batteries. Ag nanoparticles can improve electrical conductivity of the O_2 electrode and provide a catalytic activity for ORR. Mesoporous Co_3O_4 nanosheets with a thickness of 10 nm are beneficial to accommodating insoluble discharge products and facilitating oxygen diffusion and electrolyte impregnation (Figure 1.2g). Due to the synergistic effect, Ag/M-Co_3O_4/NF exhibited a robust catalytic activity, delivering a high reversible capacity of 2471.1 mAh g^{-1}, a low discharge/charge overpotential, and a cycle life over 124 cycles (Figure 1.2h). The RuO_x functionalized In_2O_3 coated freestanding carbon paper was designed by Kim and coworkers (Figure 1.2i) [28]; the In_2O_3 is unreactive to superoxides and can prevent direct contact between the carbon paper and the Li_2O_2 to generate parasitic side products such as Li_2CO_3 (Figure 1.2j). The catalytic RuO_x leads to a significant reduction in the overpotential (Figure 1.2k), acting as an electron transfer bridge between the Li_2O_2 and In_2O_3 layer. Thus, the cycling stability of Li–O_2 batteries can be extended to 165 cycles. For the aforementioned cathodes, either nickel foams or carbon paper/cloths with a diameter over 10 μm are used as the current collector, which has low surface area and limits the contribution to capacity. To increase the surface area, freestanding nanofibers with small diameters and rich active sites were prepared by the electrospinning method. Freestanding activated carbon nanofibers (ACNF) were prepared through electrospinning combined with CO_2 activation by Zhang's group [32]. The loosely packed nanofibers can improve the permeability of O_2 across the cathode. The mesopores introduced by CO_2 activation act as additional nucleation sites for Li_2O_2 deposition, leading to reduced size and flake-like individual Li_2O_2 particles.

Apart from the carbon-based cathodes, carbon-free cathodes have been investigated because of their chemical stability [33]. Carbon materials can react with superoxides to produce Li_2CO_3, exhibiting unstable properties. Carbon-free cathodes can prevent this side reaction to improve the stability of cathodes in a long cycling process. A yolk–shell catalyst of Co_3O_4@$NiCo_2O_4$ was prepared using ZIF-8@ZIF-67 materials as the template and then coating with $NiCo_2O_4$ (Figure 1.3a) [34]. Co_3O_4@$NiCo_2O_4$ composite exhibited excellent ORR and OER performance due to the adjustable structure, abundant active sites and open space (Figure 1.3b), delivering a high discharge capacity of 11 672.8 mAh g^{-1} and a long cycle life of 280 cycles. Compared with metal oxides, all metal-based matrixes have received much attention because of their high conductivity. Recently, Xu and coworkers nanoengineered Au-coated Cu nanoneedle arrays grown directly on a Cu foam structure (Au/Cu@FCu) as both the anode backbone and the cathode in a Li–O_2 battery [35]. Due to their excellent conductivity, high porosity, large specific surface, and superior lithiophilicity as well as high catalytic activity, Au/Cu@FCu electrodes can simultaneously regulate uniform deposition of lithium metal on

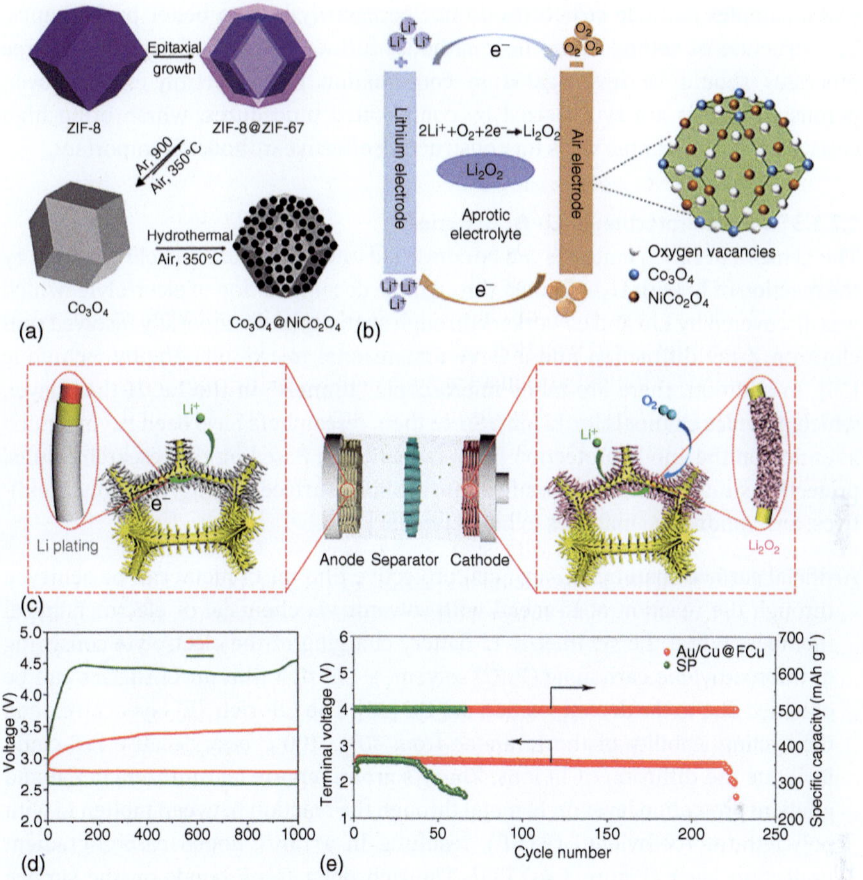

Figure 1.3 (a) Preparation of $Co_3O_4@NiCo_2O_4$; (b) electrochemical mechanism of the cells with $Co_3O_4@NiCo_2O_4$ as catalyst. Source: Reproduced with permission Zhao et al. [34]. (c) Synthetic mechanism of Au/Cu@FCu for the symmetrical Li–O_2 battery; (d) first discharge–charge voltage profiles of the Li–O_2 battery with the SP cathode and Au/Cu@FCu cathode at a current rate of 100 mA g^{-1} with a limited capacity of 1000 mAh g^{-1}. (e) Discharge terminal voltage vs. cycle number for two different cathodes with 100 mA g^{-1} at a fixed capacity of 500 mAh g^{-1}. Source: Reproduced with permission Luo et al. [35].

the anode and catalyze the formation/decomposition of Li_2O_2 on the cathode (Figure 1.3c). As a result, the Au/Cu@FCu cathode demonstrated extremely low overpotentials (0.64 V) (Figure 1.3d) and a very stable cycling stability over 220 cycles with 100 mA g^{-1} at a fixed capacity of 500 mA g^{-1-1} (Figure 1.3e). However, until now, the number of reported metal-based cathodes is less than that of other metal oxides or carbonaceous cathode materials due to their relatively low surface area and high mass density. Thus, designing suitable nanostructured metals is vital to improve their application.

Based on the preceding discussion, we know that materials with various structures and components have been adopted in Li–O_2 batteries. However, in some cases, complex cathode structures do not necessarily lead to better performance. The structure of cathodes and how cathodes influence the discharge and charge processes should be investigated in coordination with electrolytes. Moreover, porous structures are synthesized by complicated procedures, which bring high costs. Developing simple ways for constructing effective cathodes is important.

1.2.1.3 Anode Protection of Li–O_2 Batteries

The Li metal in Li–O_2 batteries can be converted into LiOH during cycling caused by the reaction of Li with H_2O formed through the decomposition of electrolyte, which was discovered by Liu and coworkers through spatially and temporally resolved synchrotron X-ray diffraction and a three-dimensional micro-tomography technique [36]. In addition, there are many microscopic "tunnels" in the LiOH thick layer, which provides channels for Li ions. Since then, researchers have been paying much attention on the anode protection in Li–O_2 batteries. Based on the research results, protection strategies can be classified into artificial surface coating, electrolyte additives, and conductive matrixes to host Li metal [37].

Artificial surface coating: An artificial protective film on Li metal can be achieved through the reaction of Li metal with solvents via chemical or electrochemical methods. When the symmetric Li battery charging in the electrolyte consisting of fluoroethylene carbonate (FEC) solvent, a LiF-rich film on Li surface can be obtained due to the decomposition of FEC [38]. The LiF-rich SEI layer can extend the cycling stability of the Li anode from 40 to 100 cycles, because LiF could facilitate the diffusion of Li ions. Zhang's group reported a multifunctional and gradient protection layer on Li metal through the reaction between molten Li with poly(tetrafluoroethylene) (PTFE), resulting in a LiF/F-doped carbon-gradient protection layer (Figure 1.4a) [39]. The rich polar C—F bonds on the surface could uniformly capture Li ions and regulate the electron configuration of LiF to make Li ions quasi-spontaneously diffuse from carbon to LiF surface. The LiF at the bottom could make fast Li ions transportation and homogenize the nucleation sites on lithium. As a result, this well-designed gradient protection layer offers dendrite-free Li metal anodes during cycling process in both ether-based and carbonate ester-based electrolytes [39]. Recently, molecular layer deposition was also reported by Adair to form a protective zircon layer on Li metal, resulting in a long life of 500 cycles in Li–O_2 batteries (Figure 1.4b) [40]. The artificial SEI layer via electrochemical or chemical methods are nanoscaled irrespective of thickness or structures, which may be destroyed during the long cycling process. To improve the stability, thick and mechanically stable films need to be designed on Li metal. Luo and coworkers designed a thick SiO_2/GO (graphene oxide) hybrid film (~22.0 μm) on Li metal surface (Figure 1.4c) [41]. The GO sheets act as a barrier to prevent the attack from oxygen species and moisture and suppress the growth of Li dendrites; the SiO_2 nanoparticles embedded on the GO sheets can effectively inhibit the agglomeration of GO sheets, as well as provide rich channels for Li ion transportation. The Li–O_2 battery with thick SiO_2/GO film-protected Li anode can stably cycle for 348 times with a capacity of 1000 mAh g^{-1}. Although artificial surface coating can obtain good cycling stability and effectively suppress the dendrite growth, the coating layer normally could increase the interfacial resistance of Li–O_2 batteries, leading to high overpotentials and fading of cycling performance.

Electrolyte additives: Exploring suitable electrolyte additives to in situ form a functional SEI protective layer during cycling process is also regarded as a promising strategy [44]. Fox example, Zhang's group added tetraethyl orthosilicate (TEOS) into the electrolyte [42]. TEOS can spontaneously react with the main lithium corrosion product (LiOH) to form a Si-O-containing film on the lithium metal anode and provide a dynamic self-healing effect. This film could effectively prevent lithium from further corrosion by O_2, H_2O, and discharge intermediates in the electrolyte (Figure 1.4d). Grey and coworkers added LiI into the electrolyte of Li–O_2 batteries with porous rGO as cathode [43]. Due to the presence of LiI, the Li–O_2 battery operation involves the reversible formation and removal of LiOH crystals instead of Li_2O_2 (Figure 1.4e), which can tolerate large quantities of water. Recently, Zhang's group developed a solid-like electrolyte by adding hydrophobic silica into the electrolyte [45]. Due to the electrostatic interaction between $CF_3SO_3^-$ and silica particles, viscosity of the optimized electrolyte was increased by 980-fold; thus diffusion of H_2O or other contaminants was much slower, which can protect the Li metal anode. The Li anode in this Li–O_2 battery achieved a long life of 550 cycles.

High-concentration electrolytes: High-concentration electrolytes (HCEs) were also studied to protect Li metal anodes through changing the solvation structure of Li ions and optimizing the contents or structures of the SEI layer. Regulating the species and concentration of salts in electrolytes can optimize the Li^+ solvation structure and adjust the SEI film to strengthen Li anode. Zhang's group adopted three salt concentrations of LiTFSI/DME (1,2-dimethoxyethane) electrolyte (1, 2, and 3 M) in Li–O_2 batteries [46]. The HCE (3 M) can lead to well-protected Li metal anode and air electrode, resulting in high capacity and stability, because the HCE can enhance the stability of electrolyte against reduced oxygen intermediates.

Metal alloys: Metal alloys instead of pure Li metal were also applied to suppress the growth of Li dendrites, such as Li/Si [47] and Li/Al [48]. But the energy density is undermined because exotic metals cannot deliver energy. Moreover, the alloying procedure is tedious and complicated.

In this section, the reaction mechanism and strategies for cathode design and anode protection of Li–O_2 batteries are discussed. Although some challenges have been successfully resolved in the past decade with various approaches, to further promote the development of Li–O_2 batteries, great efforts should be made, and several directions are provided. The mechanism of Li_2O_2 decomposition on a cathode catalyst is complicated and needs intensive investigations, especially for finding effective cathode materials with high OER catalytic activity. For practical applications of Li–O_2 batteries, the effect of CO_2 in air should be considered, which can react with Li_2O_2 to produce Li_2CO_3. In addition, as the decomposition potential of

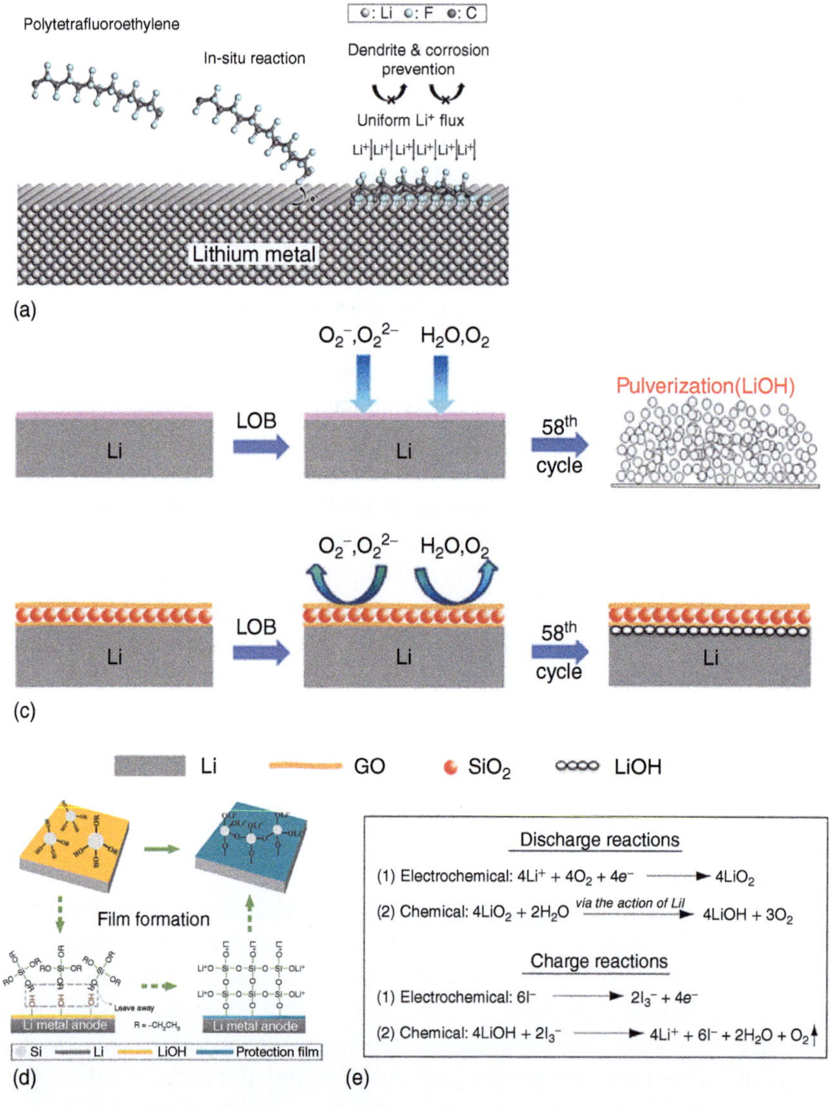

Figure 1.4 (a) Schematic illustration of the design guideline of PTL. Source: Reproduced with permission Yu et al. [39]. (b) Illustration of a protective zircon coating on a Li metal anode. Source: Reproduced with permission Adair et al. [40]. (c) Schematic illustration of mechanisms of SiO_2/GO hybrid thin layer for uniform Li deposition. Source: Luo et al. [41]. Licensed under CC BY 4.0. (d) Schematic illustration of the lithium protection film formation mechanism on TAL in Li–O_2 batteries. Source: Reproduced with permission Yu et al. [42]. (e) Schematic illustration of mechanisms for the formation and removal of LiOH in iodide redox-mediated Li–O_2 cells in the presence of water. Source: Reproduced with permission Liu et al. [43].

Li_2O_2 and Li_2CO_3 is high, it is necessary to find catalysts to decompose them at a low potential (<3.5 V). Another problem is related to the Li metal anode, such as the growth of Li dendrites. Solid-state Li–O_2 batteries should be paid much more attention, such as the improvement of conductivity, the interfacial properties of electrode and electrolyte. If these challenges can be solved, Li–O_2 batteries can take great strides forward.

1.2.2 Lithium–Sulfur Batteries

Compared with the traditional Li-ion batteries based on insertion/extraction mechanism, the lithium–sulfur (Li–S) batteries (Figure 1.5a), based on the redox reaction between Li metal and S element (Figure 1.5b), are regarded as one of the most promising next-generation energy storage systems owing to their high specific capacity (1675 mAh g^{-1}) and energy density (2600 Wh kg^{-1}), as well as the abundance of sulfur on the earth [49, 50]. Nevertheless, practical applications of Li–S batteries are hindered by intrinsic problems such as the insulating nature of sulfur and its discharge products (Li_2S), high soluble polysulfides as intermediates

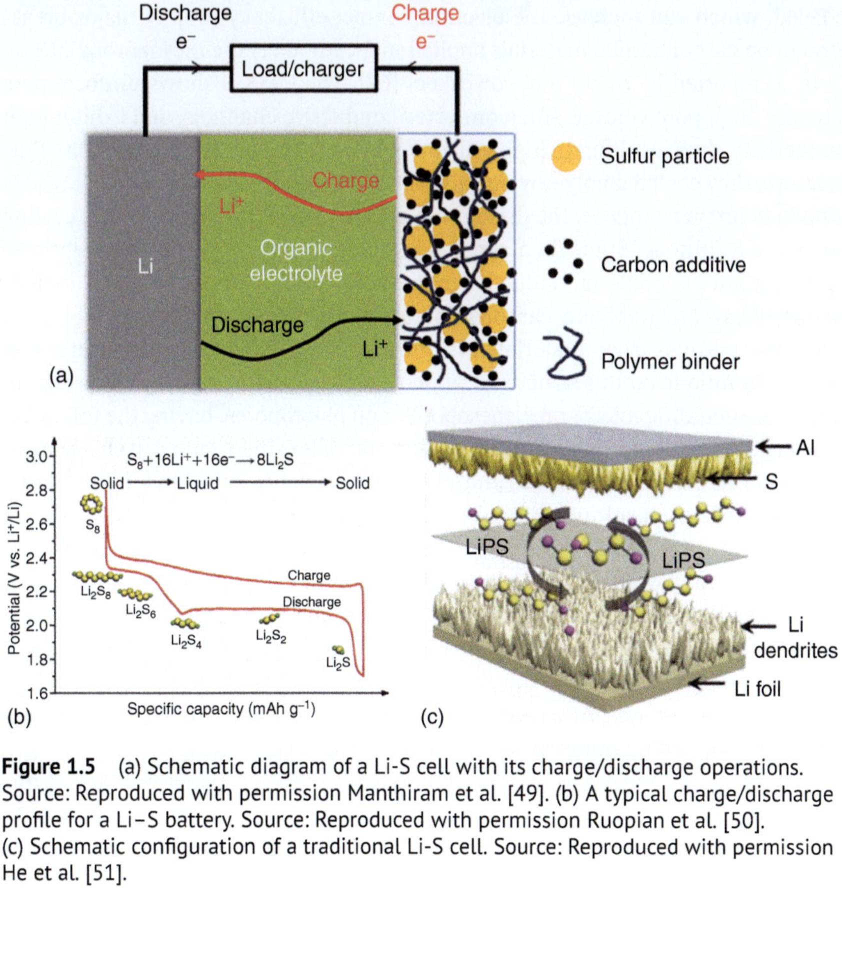

Figure 1.5 (a) Schematic diagram of a Li-S cell with its charge/discharge operations. Source: Reproduced with permission Manthiram et al. [49]. (b) A typical charge/discharge profile for a Li–S battery. Source: Reproduced with permission Ruopian et al. [50]. (c) Schematic configuration of a traditional Li-S cell. Source: Reproduced with permission He et al. [51].

during charge/discharge processes, large volume changes of electrodes, and growth of Li dendrites on the Li metal anode (Figure 1.5c) [51].

Since reported in the 1960s [52], Li–S did not get much attention due to the unclear reaction mechanism of sulfur in different electrolytes [53–55], until the cathode of sulfur embedded into a polyacrylonitrile (PAN) matrix was designed in a carbonate-based electrolyte in 2002, which delivered a relatively high and stable cycling performance (above 600 mAh g^{-1} after 50 cycles) [56]. Especially, in the past two decades, flourishing developments have been made in Li–S batteries to further enhance their electrochemical performance to promote their commercial application, including designing numerous conductive matrixes for S, introducing catalysts to promote the conversion kinetics, modifying separators to suppress the shuttle effect, exploring suitable electrolytes, and inhibiting Li dendrites. As so far, considering many review papers focused on sulfur cathode [57, 58], separators [59], and electrolytes [60] have been published, in this chapter, we mainly outline the initial and key achievements for the development of Li–S technology.

1.2.2.1 Conductive Matrixes for S Cathode

Carbon materials were first reported as conductive matrixes for sulfur in 1989 by E. Peled, which can increase the electron contact efficiency [61]. A major breakthrough on carbonaceous materials applied in S cathode is the mesoporous carbon (CMK-3) reported by Nazar and coworkers [62]. The CMK-3 shows uniform pore structure, high pore volume, interconnected conductive channels, and exhibit high conductivity, thus resulting in high utilization of active materials (Figure 1.6a). Furthermore, they coated a polyethylene glycol (PEG) layer on the surface of CMK-3/S cathode to further suppress the dissolution of polysulfides and improve the cycling stability of S cathode. Since then, great achievements have been made to fabricate porous carbon materials as conductive matrixes for sulfur cathodes [66, 67], including porous carbon, graphene, carbon nanotubes, and carbon nanofibers.

Recently, porous carbon materials as sulfur matrixes have received huge attention due to their unique porous structures, which usually possess different kinds of pore structures, including micropores, mesopores, and macropores, having the following advantages: (i) microporous pores with a diameter of less than 2 nm can encapsulate small S molecules; (ii) mesopores and macropores usually have large pore volume to loading high mass sulfur; (iii) high surface area to increase the intimate contact of sulfur with conductive matrixes, thus improving the conductivity of cathodes. In 2012, Wan's group designed a microporous carbon matrix with a pore size of only 0.5 nm, which can only confine small S_{2-4} molecules [63]. Therefore, the transition between S_8 to soluble polysulfides (Li_2S_n, $n = 4$–8) was prevented (Figure 1.6b), which can essentially solve the shuttle effect of polysulfides and improve the cycling stability of Li–S batteries. As a result, the as-obtained S_{2-4} small molecules confined in microporous carbon materials delivered a high specific capacity of 1670 mAh g^{-1} and a capacity of 1149 mAh g^{-1} after 200 cycles. To further take advantage of different pores, micro-/meso-/macroporous carbon materials with three kinds of pore properties were prepared by Tang and coworkers [68]. They systematically studied the effect of different pores on the electrochemical performance of sulfur cathodes

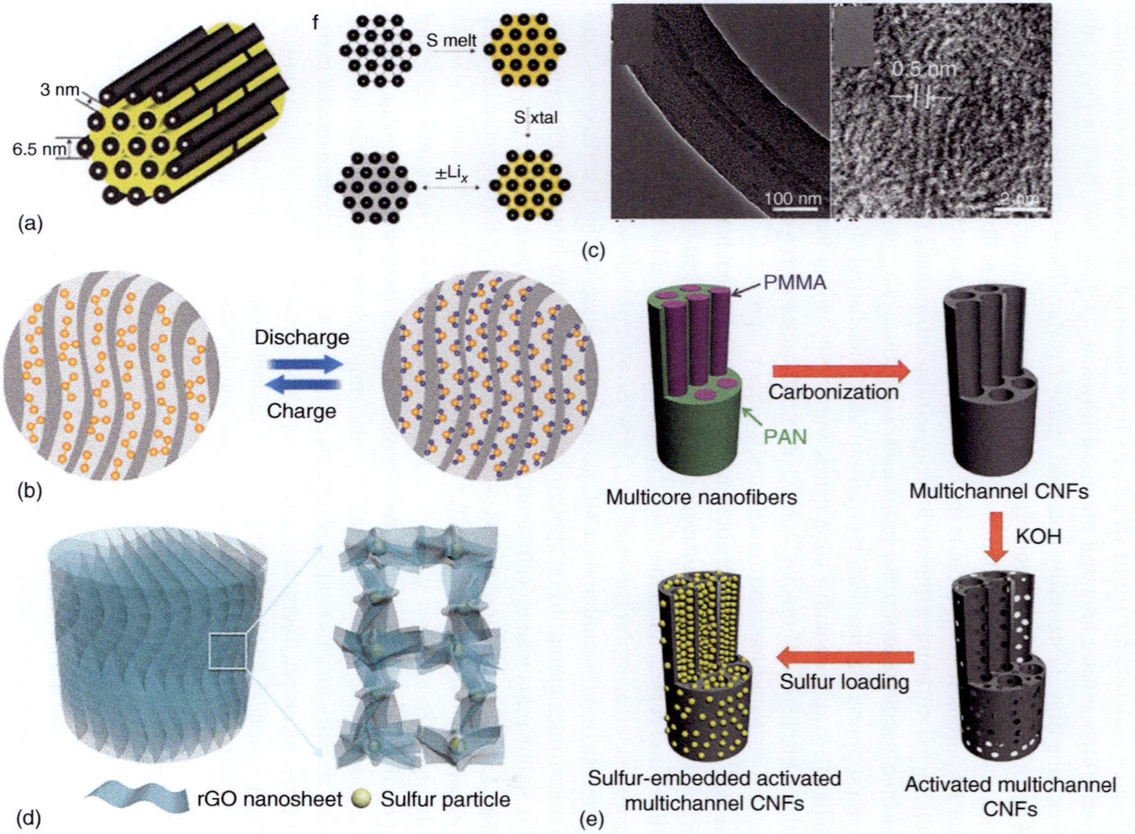

Figure 1.6 (a) A schematic diagram of S/CMK-3 composites in the discharging–charging with Li. Source: Reproduced with permission Xiulei et al. [62]. (b) Structural characterizations of CNT@MPC. (c) TEM image of a CNT@MPC nanocable. Source: Xin et al. [63] / with permission of American Chemical Society. (d) Schematic of freestanding 3D graphene/sulfur composite. Source: Reproduced with permission Papandrea et al. [64]. (e) Illustration of the sequential fabrication steps for the sulfur-embedded activated multichannel carbon nanofibers. Source: Reproduced with permission Lee et al. [65].

and reported that the pore size distribution instead of specific surface area and total pore volume influence the performance of Li–S batteries. Furthermore, the ratio of micropore volume dominates the cycling stability of S cathodes. This work will give some instructions for designing porous carbon materials for Li–S batteries.

Graphene, as a 2D carbon material, is regarded as one of the most promising conductive matrixes for sulfur due to its high conductivity, excellent mechanical property, high surface area and chemical/physical stability [69]. As the graphene freestanding film can be easily obtained, many freestanding graphene–sulfur cathodes for Li–S batteries have been reported. For example, a freestanding 3D graphene framework was prepared by Papandrea and coworkers as sulfur matrixes [64]. Due to the 3D porous structure, the mass loading of sulfur can reach as high as 90 wt%, which can greatly improve the specific capacity based on mass. In addition, the interconnected graphene walls offer many effective pathways for electron transportation and can suppress the dissolution of polysulfides into electrolyte (Figure 1.6c). Therefore, the 3D graphene/sulfur composites with 90 wt% sulfur loading could deliver a specific capacity of 341 mAh g^{-1} at 1 C after 500 cycles with a capacity retention of 77%. Carbon nanotubes and nanofibers, as other representative carbon materials, were also attracted much attention as skeletons for sulfur loading owing to their high conductivity and unique fabric structures. As a typical example, an activated multichannel carbon nanofiber (a-MCNF) was synthesized by Lee and their groups via a simple electrospinning method, followed by carbonization and KOH activation process [65]. The obtained a-MCNF materials possessed high surface area and rich micro-/mesopores. When applied as sulfur matrix, the large mesopores ensure a high sulfur mass loading (80%) and micropores help to trap soluble polysulfides (Figure 1.6d). The prepared S-a-MCNF cathode can deliver a high specific capacity of 1100 mAh g^{-1} even at a high sulfur loading of 4.6 mg cm^{-2}.

Although carbonaceous materials as sulfur host have yielded relatively stable and long cycling performance, pure carbon materials can only physically trap soluble polysulfides, which is not enough for realizing high specific capacity and long cycling stability for practical applications. Heteroatom-doped carbonaceous materials were developed because heteroatoms could strongly anchor the polysulfides through chemical bonds [70–73]. N- and O-doped carbon materials can achieve better electrochemical performance than that of other element-doped carbon materials, such as B, F, P, S, Cl, reported by Zhang's group [74]. To speed up the conversion process of soluble polysulfides, electrocatalysts were also introduced in the matrixes for sulfur [75–79], including metal oxides [80], metal sulfides [81], metal carbides [82], metal nitrides [83], metal phosphides [84], metal selenides [85] as well as single atoms [86, 87]. Apart from these metal-based catalysts, metal-free catalysts should be paid more attention, including carbon nitride, black phosphorus, and boron nitride. The mechanism of catalysts is complicated because the conversion reaction of sulfur with Li metal usually consists of four different stages: solid–liquid process, liquid–liquid process, liquid–solid process, and solid–solid process (Figure 1.7) [76]. Moreover, different polysulfides show different solubilities and physical properties in different electrolytes. Therefore, comprehensive understanding of how catalysts work is essential to guide the further designing of

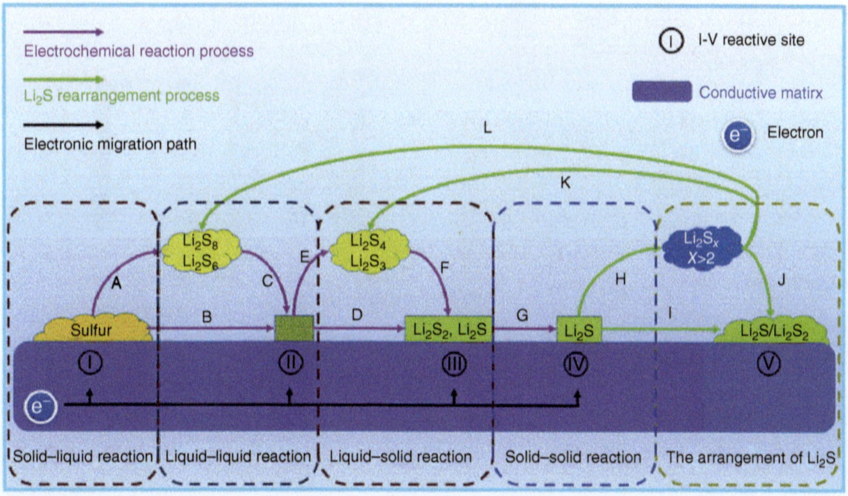

Figure 1.7 The electrochemical reaction process of the sulfur electrode. Source: Reproduced with permission Deng et al. [76].

suitable catalysts to achieved fast conversion kinetics and long-life Li–S batteries. The binding energy of different catalysts with Li_2S_6 was calculated by Y. Cui's group [88], indicating higher binding energy can lead to a better anchoring effect, thereby to mitigate polysulfide dissolution and suppress the shuttle effect. However, higher binding energies do not guarantee faster conversion kinetics because if the binding energy is too high to prevent the contact between polysulfides with conductive matrixes, it will lead to lower specific capacity and fast capacity decay.

1.2.2.2 Modifying Separators of Li–S Batteries

Separator is also one of the most components in Li–S batteries, which can directly prevent direct contact with cathode and anode, but allow the transportation of ions. Commercial separators used in Li–S batteries are polyolefin porous membranes, such as polyethylene (PE), polypropylene (PP), or multilayer composite membranes (PEP), which all have a pore size of about 100 nm. Big pores allow soluble polysulfides to pass through, leading to the shuttle effect. The modified separators can keep the dissolved polysulfides into the cathode side via the physical or chemical adsorption mechanism to improve the utilization of active materials and enhance the long-term cycling stability [89]. Manthiram's group first reported a microporous carbon film as an interlayer between the cathode and separator to block soluble polysulfides into the cathode area and suppress the shuttle effect (Figure 1.8a) [90]. In addition, the interlayer can be regarded as a current collector to improve the utilization of active materials. Later, they directly coated ultralight multiwall carbon nanotube (MWCNT) layers on the commercial PP separator. The MWCNT layers could effectively trap polysulfides and improve the cycling stability.

Similarly to the cathode matrixes, separators decorated by materials having strong chemical bonds with polysulfides were also reported. Separators coated with lightweight B- and N-doped reduced graphene oxides (rGO) were also prepared

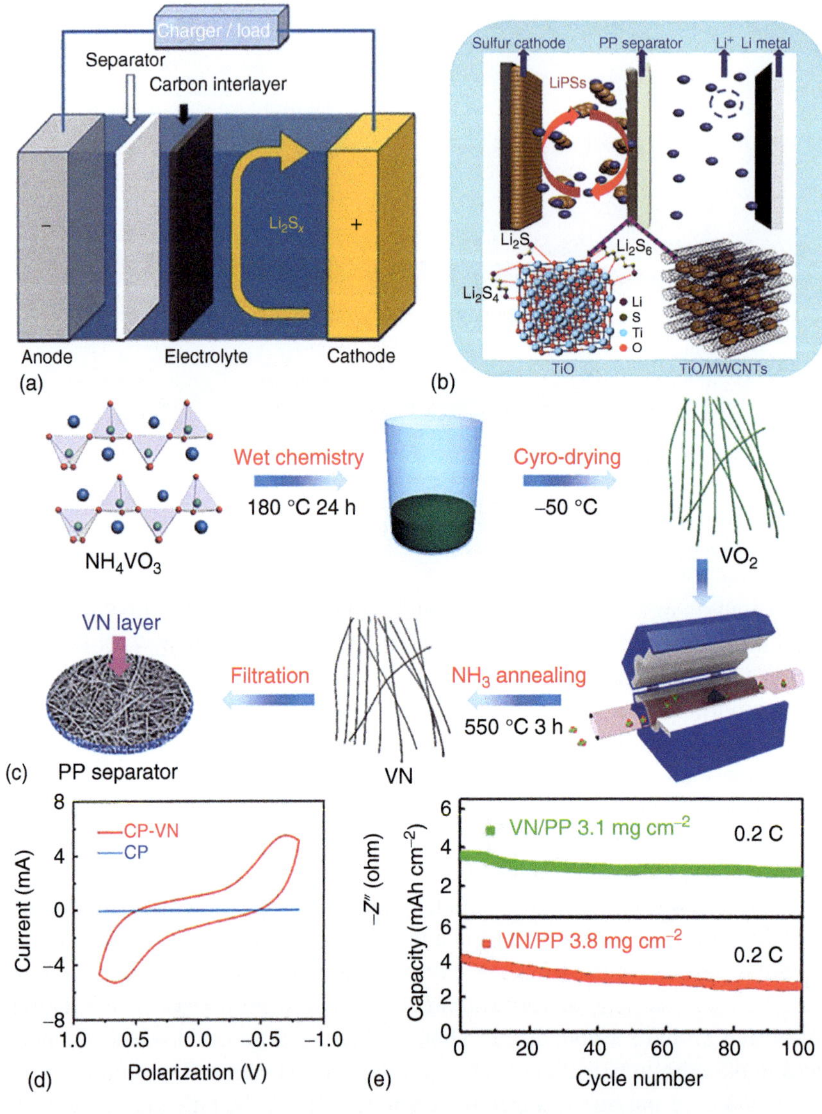

Figure 1.8 (a) Schematic configuration of a Li–S cell with a bifunctional microporous carbon interlayer. Source: Reproduced with permission Su et al. [90]. (b) Schematic configuration of TiO/MWCNTs coated separator for suppressing the shuttle effect. Source: Reproduced with permission Li et al. [91]. (c) Schematic showing the fabrication process of the VN-modified separator; (d) CV of symmetric cells using CP–VN and CP as electrodes; (e) cycling stabilities of S@VN/PP with sulfur mass loadings of 3.1 and 3.8 mg cm^{-2} at 0.2 C. Source: Song et al. [92] / with permission of American Chemical Society.

as a polysulfide reservoir [93]. The heteroatoms on rGO could create polar carbon lattices to form strong interaction with polysulfides, thus suppressing the shuttle effect. The commercial titanium monoxide (TiO) and MWCNTs coated separator was designed by Wang and coworkers (Figure 1.8b), which is a simple and low-cost process [91]. Compared with TiO_2, TiO offers higher conductivity, and based on the DFT calculation results, TiO has better adsorption effect. The Li–S batteries based on TiO/MWCNTs coated separator delivered a high capacity of 1033.8 mAh g^{-1} at 0.5 C with an ultralow fading rate of 0.057% per cycle. Metal-based materials were also used to decorate separators, due to their strong interaction with polysulfides as well as the catalytic effect they offered. VN-membrane-based separator was prepared by Sun and coworkers (Figure 1.8c) [92]. Due to excellent electronic conductivity of VN ($\sim 1 \times 10^6$ S m^{-1}), the VN-coated separator not only can act as a chemical reservoir to trap polysulfides but also can promote the redox kinetics (Figure 1.8d). Thus, Li–S batteries with VN-based separator delivered a low decay rate about 0.077% per cycle at 1 C for 800 cycles. A high areal capacity of 4.2 mAh cm^{-2} at 0.2 C can be obtained even with a high mass loading of 3.8 mg cm^{-2} (Figure 1.8e), indicating high sulfur utilization because of effective polysulfides trapping and fast redox kinetics induced by VN.

Modified separators can indeed suppress the shuttle effect and improve the cycling stability. However, multifunctional modified separators should be paid more attention. For example, materials used to modify separators could not only chemically adsorb polysulfides, but also can promote the conversion kinetics of polysulfides. In addition, modified separators will not affect the transportation of ions or increase the interface resistance. All these factors should be considered when designing suitable materials to decorate separators.

1.2.2.3 Electrolyte Design for Li–S Batteries

The electrolyte is one of important parts in batteries, which should have good ionic conductivity. In Li–S batteries, electrolyte is more important because the discharge intermediates are soluble in ether-based electrolytes (1,3-dioxolane [DOL]/DME based electrolytes), which are currently the best electrolyte systems in Li–S batteries with high ionic conductivity and suitable polysulfides dissolution. The challenges are to inhibit the shuttle effect and minimize the effect of polysulfides on Li metal anodes. Controlling the solubility of solvents is the most direct and effective method, such as choosing suitable solvents or adding cosolvents to optimize the solubility of polysulfides [94]. Among all kinds of solvents, fluorinated ethers solvents are widely applied in Li–S batteries as cosolvents because they can improve the viscosity of electrolytes and improve the mass-transfer process [95]. In addition, chelating oxygen by fluorine could change the solvation properties of fluorinated ethers, which can affect the solubility of lithium salts or polysulfides in electrolytes. Fluorinated diether (FDE) was reported by Wen's group as a cosolvent in the electrolyte of 1 M LiTFSI in DOL/DME [95]. Due to the insolubility of polysulfides in FDE, dissolution of polysulfides in the FDE modified electrolyte was suppressed. In addition, the corrosion of the Li metal anode was also decreased. As a result, the Li–S batteries with this optimized electrolyte exhibited a high reversible capacity of

701 mAh g^{-1} at 0.5 C after 200 cycles with an average coulombic efficiency of 99% (Figure 1.9a–c).

Another promising strategy is to increase the Li salt concentration. Because in the concentrated electrolytes, the solvents will form contact ion pairs (CIPs) or aggregates (AGGs) and the free solvents will disappear, resulting in insolubility of polysulfides [98–100]. Many researchers have reported that when the concentration of electrolytes is up to 5 M, both dissolution and diffusion of polysulfides decrease, leading to high coulombic efficiency of over 99%. However, with the increasing concentration, the ionic conductivity decreases, and voltage polarization is more obvious. Solid-state electrolytes (SSEs) were also explored in Li–S batteries. In SSEs, S can be directly transferred to Li_2S without producing soluble polysulfides; thus the shuttle effect and dissolution of polysulfides are both eliminated. Although the cycling stability of Li–S batteries can be improved, poor rate capability and high interfacial resistance are observed at room temperature due to low ionic conductivity. To date, numerous SSEs have been reported by researchers [101–103], including mainly inorganic solid electrolytes, solid polymer electrolytes, and composite solid electrolytes. In 2017, Tu's group reported an MoS_2-doped Li_2S-P_2S_5 glass–ceramic electrolyte ($Li_7P_{2.9}S_{10.85}Mo_{0.01}$) [96], which shows a high ionic conductivity of 4.8 mS cm^{-1} at room temperature and a high specific capacity of 1020 mAh g^{-1} at 0.05 C was obtained in all solid Li–S batteries at room temperature (Figure 1.9d–g). $LiBH_4$ has also been used in SSEs due to its high ionic conductivity for Li ions (over 2 mS cm^{-1}). Unemoto and coworkers reported an all-solid-state Li–S battery based on $LiBH_4$ electrolyte at 120 °C [97]. Only one plateau appeared during the cycling process, indicating that dissolution of polysulfides is inhibited (Figure 1.9h). At 0.05 C, the Li–S batteries with $LiBH_4$ electrolyte delivered an initial discharge capacity of 1140 mAh g^{-1}, and after 45 cycles the capacity can be maintained at 730 mAh g^{-1} with coulombic efficiency nearly 100% (Figure 1.9i).

Although dissolution of polysulfides and shuttle effect were completely inhibited in SSEs, in terms of reaction kinetics and the demand for fast charging/discharging, the SSEs face many challenges, especially using high sulfur mass loading and less conductive matrixes. In order to explore optimized electrolytes for Li–S batteries, ionic conductivity, suitable solubility and diffusion of polysulfides, and stability with Li metal anodes all should be considered.

1.2.2.4 Anode Protection for Li–S Batteries

Lithium–sulfur batteries use Li metal as anode, because its chemical activity is high and Li metal can easily react with electrolyte, leading to low coulombic efficiency. In addition, the formation and growth of Li dendrites on anode surface will hinder practical applications of Li–S batteries. Due to the shuttle effect, polysulfides will also react with Li metal and corrode the Li metal anode. The protection strategies for Li metal anodes in Li–S batteries include: exploring electrolyte additives to form stable SEI layer, forming artificial SEI by coating materials on the surface of Li metal and using alloy anode [104–106].

Among the electrolyte additives, $LiNO_3$ is widely used in Li–S batteries, which expertly can improve the cycling stability of Li–S batteries. Xiong's group explained

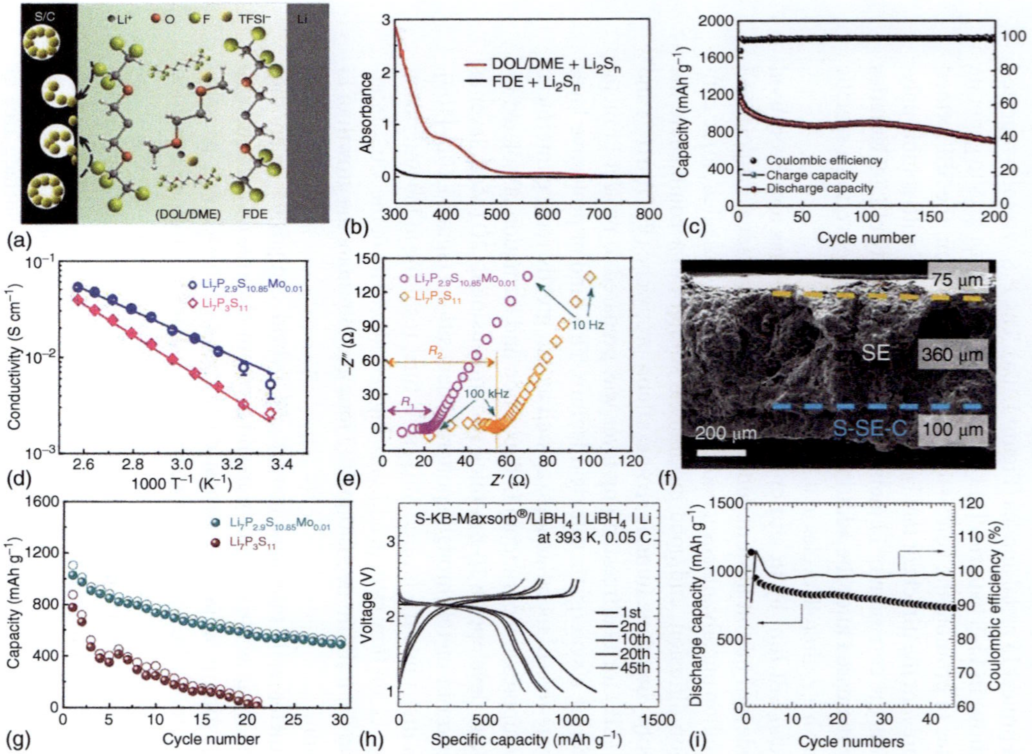

Figure 1.9 (a) Schematic diagram of fluorinated diether cosolvent electrolytes in Li–S batteries. (b) Ultraviolet–visible (UV–vis) spectra of polysulfides in different solvents; (c) cycle performance of Li–S batteries with 80% FDE electrolyte at 0.5 C. Source: Reproduced with permission Gu et al. [95]. (d) Arrhenius conductivity plots of $Li_7P_3S_{11}$ and $Li_7P_{2.9}S_{10.85}Mo_{0.01}$ at temperatures from 298 to 388 K. (e) Impedance plots of the electrolytes at room temperature; (f) cross-sectional SEM images of all-solid-state Li–S battery based on $Li_7P_{2.9}S_{10.85}Mo_{0.01}$ electrolyte; (g) cycling performances of $Li_7P_3S_{11}$ and $Li_7P_{2.9}S_{10.85}Mo_{0.01}$ as the electrolytes for all-solid-state Li–S batteries at a rate of C/20 at room temperature. Source: Xu et al. [96] / with permission of Royal Society of Chemistry. (h) Typical discharge–charge profiles of the S-KB-Maxsorb/$LiBH_4$|$LiBH_4$|Li rechargeable cell operating at 393 K and 0.05 C. (i) Discharge capacity and coulombic efficiency as a function of cycle numbers. Source: Reproduced with permission Unemoto et al. [97].

that $LiNO_3$ can react with Li metal to form a stable SEI layer [107], consisting of both inorganic species of LiN_xO_y and organic species such as ROLi and $ROCO_2Li$, which can effectively suppress the shuttle effect to protect Li metal anodes. Zhang and coworkers investigated the role of the SEI layer by *operando* the X-ray absorption spectroscopy (XAS) technique [108], indicating a dense and stable SEI layer composed of Li_2SO_3 and Li_2SO_4 formed in the first discharge process due to the synergistic effect of shuttled polysulfides and $LiNO_3$, which can effectively suppress the shuttle effect and protect Li metal anodes (Figure 1.10a). The P_2S_5 was also reported as electrolyte additive in Li–S batteries [111], which can react with Li_2S to generate soluble composites to promote the conversion of Li_2S. In addition, the SEI formed on Li metal can prevent the direct contact between the electrolyte with soluble polysulfides to suppress the corrosion of Li metal. The cycling stability of Li–S batteries was dramatically increased after the addition of P_2S_5 additive, delivering a specific capacity of 900 mAh g^{-1} after 40 cycles at 0.1 C.

Forming artificial SEI film by in situ or ex situ techniques has also been used to protect Li anode and inhibit lithium dendrites [112–114]. An ideal artificial SEI should have the following advantages: good chemical stability, excellent ionic conductivity for Li ion diffusion, and good mechanical strength to accommodate the Li dendrites. A poly(vinylidene fluoride)-*co*-hexafluoropropylene (PVDF-HFP) and lithium fluoride (LiF) formed freestanding layer was prepared to protect Li metal anodes [115], offering a synergistic soft-rigid feature. This kind of protective layer can realize dendrite-free Li metal anodes, as well as can greatly reduce interfacial resistance due to suppression of random Li deposition and formation of isolated Li. Choi and coworkers developed a 10-nm-thick two-dimensional (2D) MoS_2 layer as a protective layer on Li metal through sputtering deposition (Figure 1.10b) [109]. The Li atoms can insert into the layered MoS_2 structure, resulting in low interfacial resistance and fast Li ion flow rate. In addition, high conductivity of the MoS_2 interlayer could lead to the uniform deposition of Li ions and eliminate the growth of Li dendrite nucleation.

Replacing Li metal with micro/nanostructured hybrid composites is another promising method to guide the Li deposition and accommodate the volume changes during striping/plating process. The nanosized Li metal host with rich porosity, high surface area, and connected conductive channels can lead to unique deposition behavior of Li ions and improve the safety of Li metal anodes. Cui and coworkers designed a polymeric matrix with the lithiophilic property to load molten Li metal [110], resulting in a nanofiber porous Li anode (Figure 1.10c). In addition, the polymeric matrix can trap Li into its backbones to accommodate the volume change and suppress the growth of Li dendrites (Figure 1.10d). The high surface area of porous Li metal anodes can reduce the effective current density. Owing to the above advantages, the batteries can deliver a stable cycling performance of more than 100 cycles even at a high current density of 5 mA cm^{-2} in both carbonate and ether electrolytes. Li-alloying materials have received much attention due to their good processability and excellent theoretical capacity (3495 mAh g^{-1}). It has been reported Li–B alloy indeed could change the behavior of Li deposition. For example, Zhang and coworkers used Li–B alloy as anode instead of Li metal [116]. The alloy

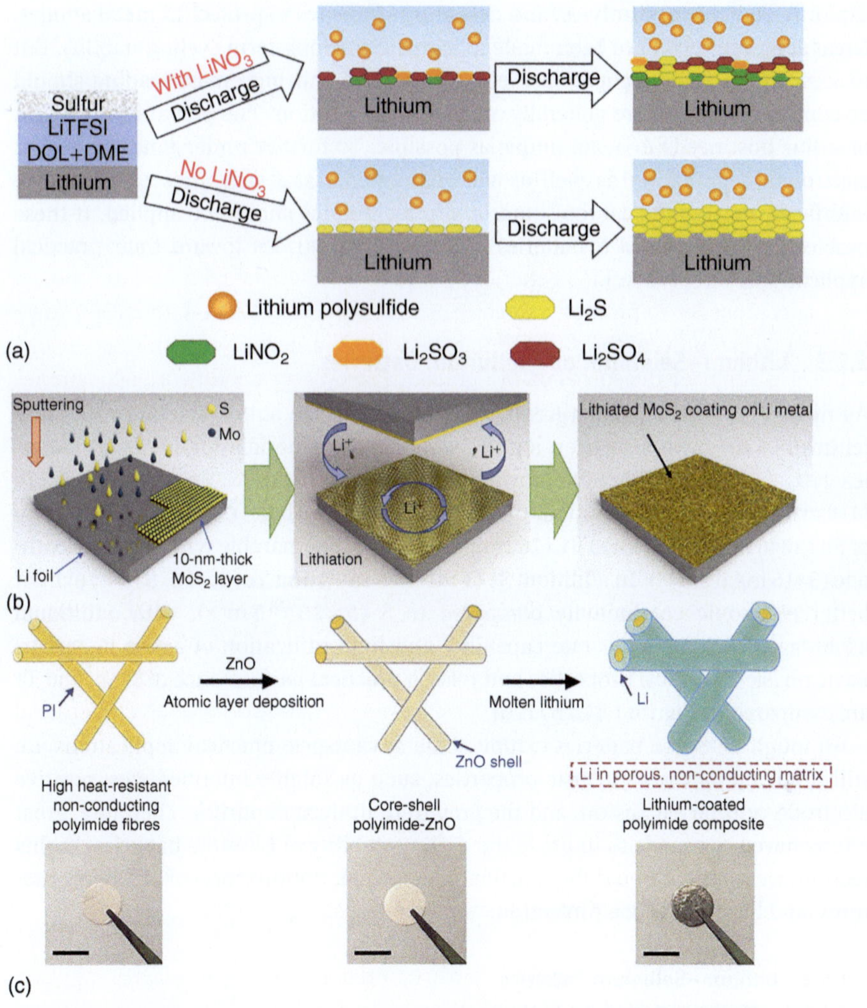

Figure 1.10 (a) Illustration of $LiNO_3$ additives for suppressing the shuttle effect. Source: Reproduced with permission Zhang et al. [108]. (b) Schematics illustrating the fabrication method for a MoS_2-coated Li anode via sputtering and subsequent lithiation. Source: Reproduced with permission Cha et al. [109]. (c) Electrospun PI was coated with a layer of ZnO via ALD to form core–shell PI-ZnO. The existence of ZnO coating renders the matrix "lithiophilic" such that molten Li can steadily infuse into the matrix. The final structure of the electrode is Li coated on a porous, nonconducting polymeric matrix. Source: Liu et al. [110] / with permission of Macmillan Publishers.

has unique structure with abundant free Li embedded in stable Li_7B_6 loofah sponge-like framework. The Li_7B_6 component could suppress the orientational deposition of Li ions, reduce effective current density, and promote the formation of stable SEI layers, resulting in stable Li–S batteries.

Until now, huge attention has been paid to Li–S batteries, including exploring conductive matrixes for sulfur, modifying separators to suppress the shuttle effect, exploring suitable electrolytes, and designing strategies to protect Li metal anodes. Great achievements have been made to increase the long-term cycling stability. But in view of practical applications, lean electrolytes and high mass loading should be considered, which are generally are paid less attention. The preparation process of sulfur host needs to be as simple as possible. To further understand anode and electrolyte degradation as well as the components and structures of electrolyte interfaces, new analytical tools and in situ techniques should be applied. If these problems are solved, Li–S batteries will take great strides toward their practical applications.

1.2.3 Lithium–Selenium or –Tellurium Batteries

As members of chalcogens (group VIA elements), like sulfur, selenium (Se) and tellurium (Te) have also been studied as cathode materials for Li metal batteries [10, 117–119]. Although gravimetric capacity of Se (675 $mAh\,g^{-1}$) and Te (419 $mAh\,g^{-1}$) is lower than that of S cathode (1675 $mAh\,g^{-1}$), gravimetric capacity of Se (3246 $mAh\,cm^{-3}$) and Te (2621 $mAh\,cm^{-3}$) is comparable with that of S cathode (3416 $mAh\,cm^{-3}$). In addition, Se ($1\times10^{-3}\,S\,m^{-1}$) and Te ($2\times10^{2}\,S\,m^{-1}$) exhibit better electronic conductivity compared to S ($5\times10^{-28}\,S\,m^{-1}$), with additional advantages such as better rate capability and high utilization of active materials. Basic physicochemical properties and related practical performance of S, Se, and Te are compared in Figure 1.11a,b [120].

Although Li–Se/Te batteries exhibit some advantages, practical applications are still hindered by their intrinsic properties, such as soluble intermediates, positive electrode volume expansion, and the growth of lithium dendrites. Therefore, great efforts have been made to improve the cycling stability of Li–Se/Te batteries. In this section, we mainly discuss the reaction mechanism, optimization of cathode structures, and Li metal anode protection.

1.2.3.1 Lithium–Selenium Batteries

1.2.3.1.1 Mechanism of Li–Se Batteries

Amine and coworkers initially reported Se/MWCNTs composites as cathode for Li–Se batteries in a carbonate electrolyte [121]. There is only one redox plateaus pair during discharge/charge process, indicating the direct conversion between Se and Li_2Se without any polyselenides (Figure 1.11c). This result is different from that of S cathode in ether-based electrolytes, which is a multiple reaction mechanism. Some researchers believed that insolubility of polyselenides in carbonate electrolytes is responsible for single-phase transformation [122]. However, some relevant works showed that multistep lithiation in carbonate electrolytes makes the working principle still ambiguous. For example, Zhou's group reported that Se nanowire cathodes exhibited multi-plateaus during cycling process in carbonate electrolytes [123]. After a large number of studies, it has been reported that the crystal structure of Se has a critical impact on the lithiation process, where amorphous Se appears in multistep lithiation, while trigonal Se delivers single-phase transformation [124, 125]. Similar to S, lithiation/delithiation of Se in ether-based electrolytes

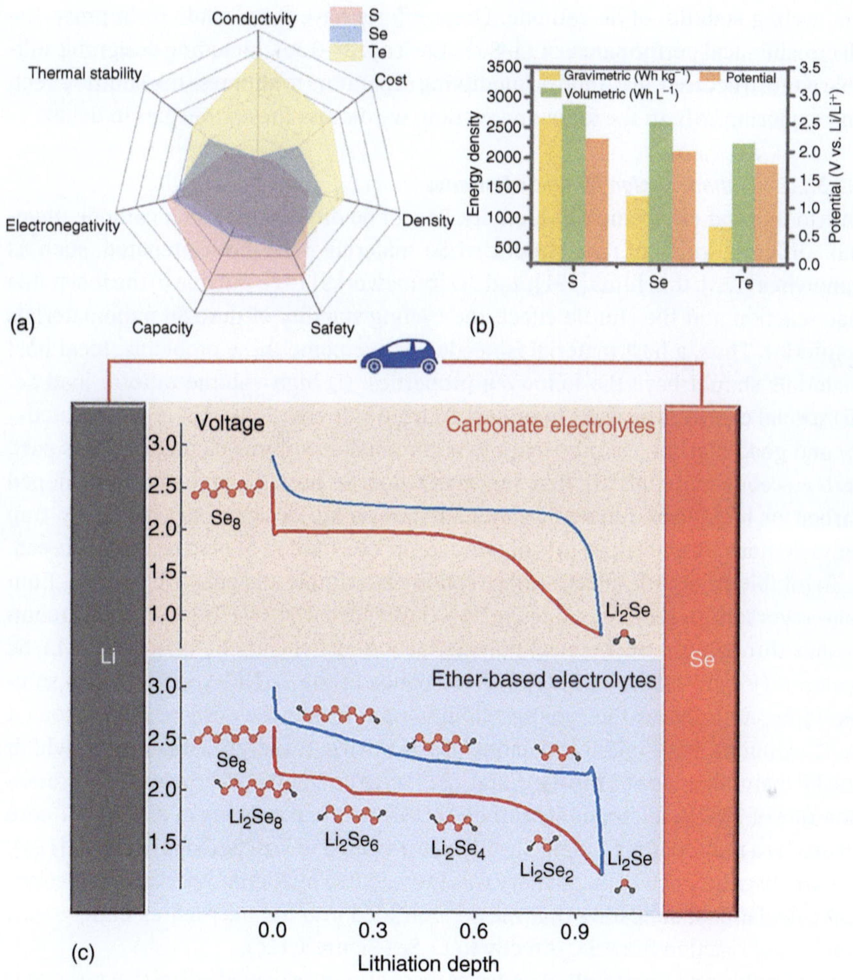

Figure 1.11 (a) Comparison of S, Se, and Te based on various properties. (b) Energy density and plateau voltage of Li–S, Li–Se, and Li–Te batteries. Source: Reproduced with permission Chen et al. [120]. (c) Schematic of the charge/discharge behaviors of Li–Se batteries in carbonate and ether-based electrolytes. Source: Reproduced with permission Sun et al. [10].

shows multiple conversion processes with a lower redox potential (Figure 1.11c) [126]. Generally, Se is first reduced to Li_2Se_n ($n \geq 4$) at ≈2.1 V, then to Li_2Se_2, and finally to Li_2Se at ≈1.9 V. While in the charge process, Li_2Se is directly oxidized to Li_2Se_n ($n \geq 4$) at ≈2.3 V by passing Li_2Se_2, and eventually to Se [127]. Polyselenides are soluble in ether-based electrolytes, which could lead to the "shuttle effect" similarly to sulfur cathode. Although the shuttle effect is not present in carbonate electrolytes, strong nucleophilic reactions between chain-like Se anions and the carbonyl groups of organic solvent will lead to severe capacity fading. Therefore, the major challenge of Li–Se batteries is to inhibit the shuttle effect and improve the cycling stability of Se cathode. Great efforts have been made to improve the electrochemical performance of Li–Se batteries [128–130], including designing suitable nanostructured Se cathodes, modifying separators to suppress the shuttle effect, and exploring SSE. In the following section, we discuss these strategies in detail.

1.2.3.1.2 Cathode Design for Li–Se Batteries

Due to its good electronic conductivity, Se can be directly used as a cathode material. Different kinds of nanostructured Se materials have been prepared, such as nanowires [123], thin films [131], and 3D frameworks [132]. But due to the inevitable side reaction and the shuttle effect, the cycling stability of pure Se nanomaterials is inferior. Thus, a host material is needed to overcome these problems. Ideal host materials should have the following properties: (i) high volume ratio to load Se; (ii) special channels or walls to protect Se from side reactions; (iii) good conductivity and good mechanical properties to accommodate volume changes. In this part, carbonaceous materials are first summarized as Se host. Then, heteroatom-doped carbon or metal-based-material-modified carbon are designed to chemically trap polyselenides. Se embedded into organic composites and Se alloys are also discussed.

To inhibit the shuttle effect, surface coating is a simple and effective method. Eom and coworkers presented conductive PANi layer coated Se/C (ET-Se/C/aniline) composites through electrochemical polymerization as the cathode material for Li–Se batteries (Figure 1.12a) [133]. The C—N bonds in the PANi layer could trap soluble Li_2Se_n to improve the cycling stability. As a result, the designed PANi coated Se/C exhibited high initial capacities of 596 mAh g^{-1} and 2873 mAh cm^{-3}, which can be maintained to 462 mAh g^{-1} and 2227 mAh cm^{-3} after 200 cycles with a retention rate of 78%. A microporous carbon (MPC) matrix consisting of a MWCNT core coated by a microporous carbon sheath was prepared to host Se (Figure 1.12b) [134]. Electrochemistry of a Li–Se_x battery was investigated by Raman spectra and theoretical calculations, indicating that the Se_x confined into the micropores undergoes a single-step reaction from Se_x directly to Li_2Se (Figure 1.12c).

The flexible and freestanding matrixes were also investigated as host materials for Se [137, 138]. Yu's group designed a porous carbon nanofibers (PCNFs)/selenium composite electrode (Se@PCNFs) by infiltrating Se into mesoporous carbon nanofibers (PCNFs) (Figure 1.12d,e) [135]. The mesopores can load high mass Se and simultaneously suppress the dissolution of active materials. When paired with Li metal, the Se@PCNFs electrode exhibits a reversible capacity of 516 mAh g^{-1} after 900 cycles without any capacity loss at 0.5 A g^{-1}. A 3D graphene-CNT@Se (3DG-CNT@Se) aerogel with carbon nanotube/selenium sandwiched between graphene nanosheets is reported by He and coworkers (Figure 1.12f) [136]. The unique 3D mesoporous, conductive network provides effective channels for electron transfer and ionic diffusion, and the hierarchical structure could suppress the dissolution of polyselenides and accommodate the volume expansion of Se during cycling. As a result, the 3DG-CNT@Se delivered an initial capacity of 632.7 mAh g^{-1} and good rate capability (high capacity of 192.9 mAh g^{-1} at 4 C).

To further improve the cycling stability of Se/C composites, carbon-based matrixes with chemical adsorption sites for polyselenides were also studied [139–141].

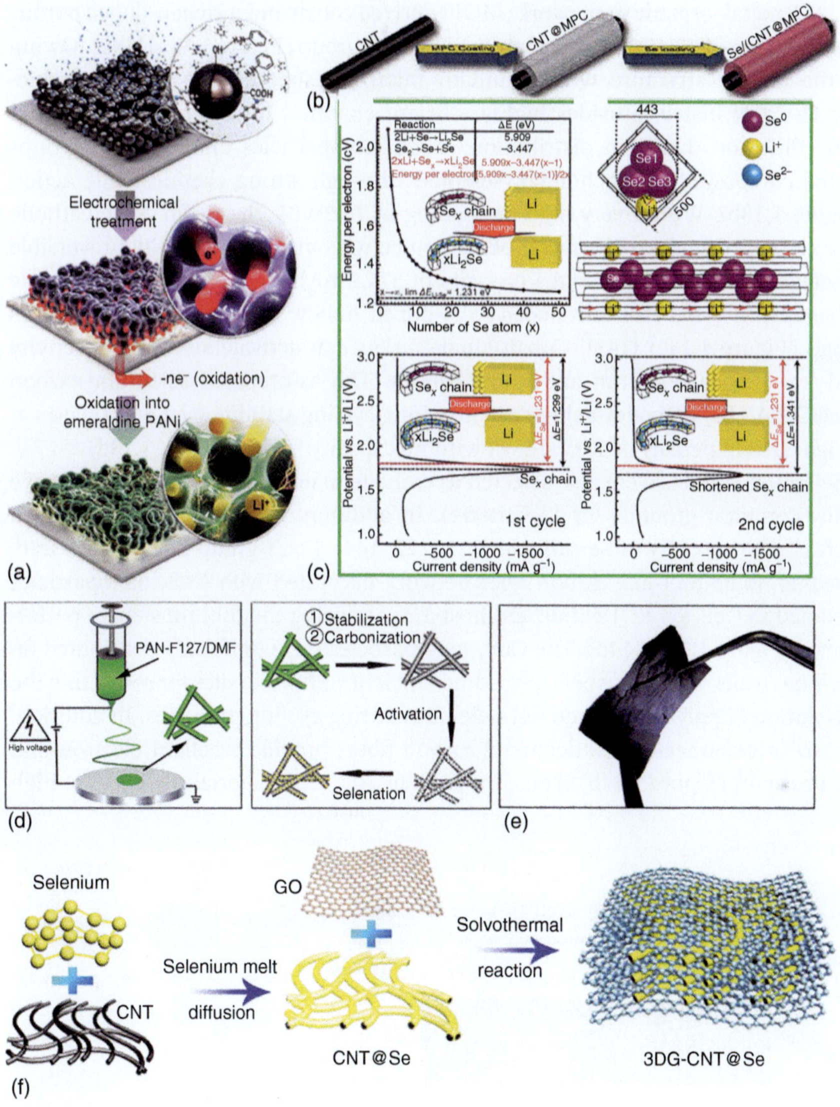

Figure 1.12 (a) Schematic illustration of the electrochemical polymerization of aniline on a Se/C cathode. Source: Reproduced with permission Lee et al. [133]. (b) Schematic diagram showing the preparation of Se/(CNT@MPC); (c) electrochemistry of a Li–Se$_x$ battery. Source: Reproduced with permission Xin et al. [134]. (d) Schematic illustration of the synthesis process and (e) photograph of the Se@PCNFs electrode. Source: Zeng et al. [135]. Reproduced with permission of (2014) John Wiley and Sons. (f) Schematic illustration of the synthesis procedure of the 3DG-CNT@Se composite. Source: Reproduced with permission He et al. [136].

In 2017, metal–organic framework (MOF) derived cobalt and nitrogen-doped porous graphitic carbon (C–Co–N) was prepared by He's group (Figure 1.13a) [142]. Owing to the unique structure with abundant micro-/mesopores, the highly conductive C–Co–N matrix provides highly efficient channels for electron transfer and ionic diffusion. Uniform distribution of Co nanoparticles and nitrogen groups in the composite can anchor polyselenides through strong chemical interaction (Figure 1.13b). With a very high Se loading of 76.5 wt%, the C–Co–N/Se cathode delivers superior electrochemical performance with an ultrahigh initial reversible capacity of 672.3 mAh g^{-1} and a capacity of 574.2 mAh g^{-1} after 200 cycles. Single cobalt atoms were also introduced to N-doped hollow porous carbon by Wang's group (Figure 1.13c) [143]. Cobalt single atoms can activate selenium reactivity and immobilize selenium and polyselenides. The as-prepared selenium–carbon (Se@CoSA-HC) cathodes deliver an ultra-long cycling stability over 5000 cycles at a high current density of 50 C almost with no capacity fading (Figure 1.13d).

Selenium sulfides were also selected as cathode materials because Se and S have many common grounds for Li batteries. In addition, introducing S into Se can increase the capacity of Se cathode [121, 127, 144]. Lou's group developed a free-standing, lotus root-like carbon fiber network decorated with CoS_2 nanoparticles (denoted as CoS_2@LRC) as the SeS_2 host for enhancing the lithium storage performance (Figure 1.14a) [145]. The CoS_2 nanoparticles were uniformly distributed on the inner walls of carbon fibers, providing efficient sulfiphilic sites for restricting the dissolution of polysulfides and polyselenides during cycling processes. In addition, the 3D interconnected multichannel carbon fibers provide excellent electron and ion transport (Figure 1.14b). Thus, the CoS_2@LRC/SeS_2 materials delivered a high

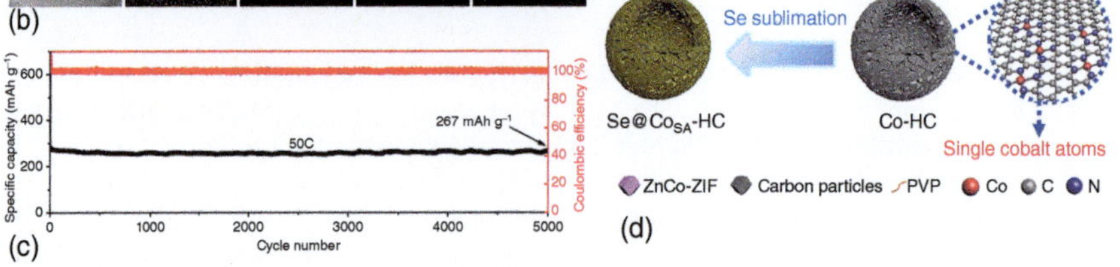

Figure 1.13 (a) Schematic illustration of the synthetic process for C–Co–N/Se; (b) the TEM image of C–Co–N/Se and the corresponding elemental mapping images of carbon, selenium, cobalt, and nitrogen. Source: He et al. [142] / with permission of Elsevier. (c) Schematic illustration of synthesizing procedures for cobalt single atoms/nitrogen-doped hollow porous carbon (CoSA-HC) particles; (d) long cycling performance and coulombic efficiency of Se@CoSA-HC at 50 C for 5000 cycles. Source: Tian et al. [143]. Licensed under CC BY 4.0.

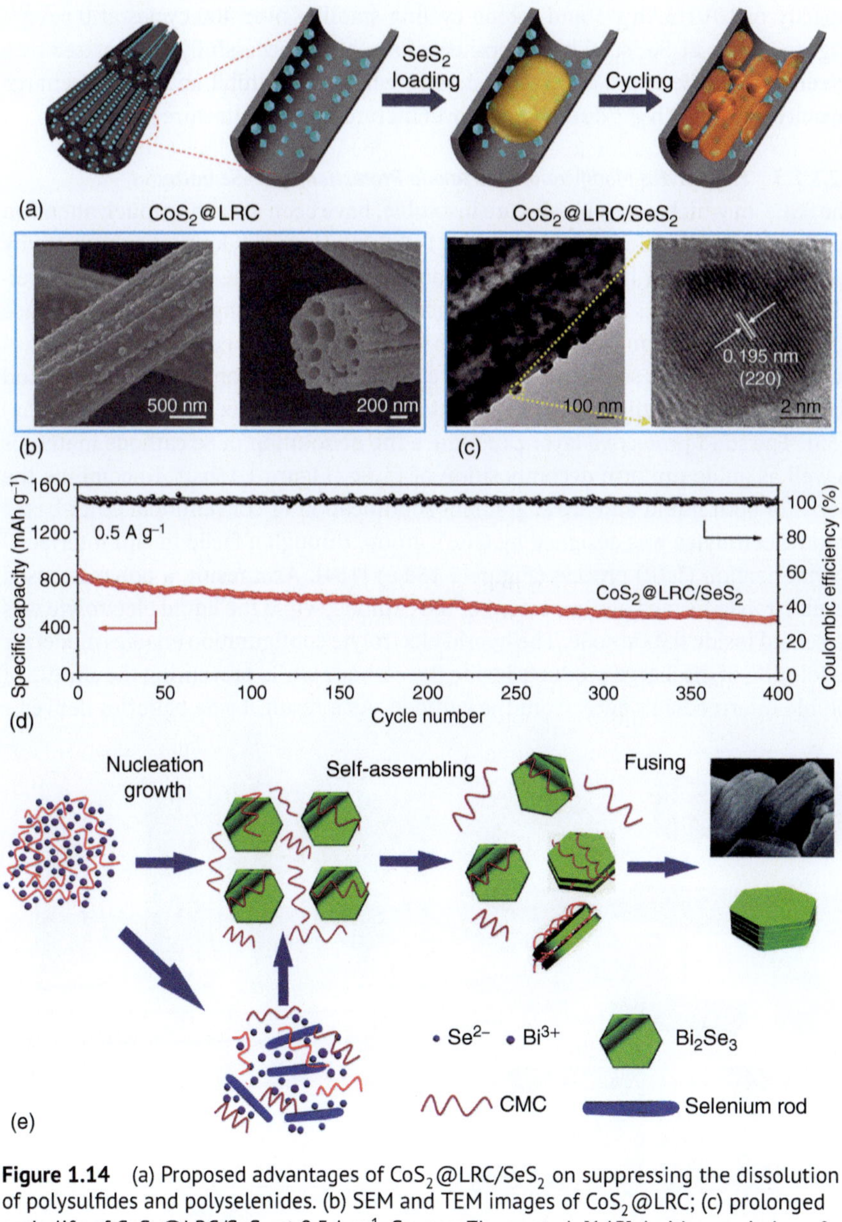

Figure 1.14 (a) Proposed advantages of CoS_2@LRC/SeS_2 on suppressing the dissolution of polysulfides and polyselenides. (b) SEM and TEM images of CoS_2@LRC; (c) prolonged cycle life of CoS_2@LRC/SeS_2 at 0.5 A g^{-1}. Source: Zhang et al. [145] / with permission of John Wiley and Sons. (d) Schematic illustration of the proposed formation mechanism of the Bi_2Se_3 stacking nanosheets. Source: Xu et al. [146]/ with permission of Royal Society of Chemistry.

capacity of 1010 mAh g^{-1} and a long cycling stability over 400 cycles at 0.5 A g^{-1} (Figure 1.14c). Bi_2Se_3 stacking nanosheets have been successfully synthesized by a "green" and facile approach (Figure 1.14d), which can exhibit an initial discharge capacity of 715 mAh g^{-1} due to their novel hierarchical architecture [146].

1.2.3.1.3 Electrolytes Modification and Anode Protection of Li–Se Batteries

The SSEs, in which polyselenides are insoluble, have been attracting much attention to solve the shuttle effect fundamentally [147–150]. However, low ionic conductivity is the main problem hindering their widespread application, as well as poor wettability between SSEs and active materials leads to increasing interface resistance [151, 152]. Exploration of SSEs for Li–Se batteries is still in its early stage; great effort should be devoted. A solid-state protective layer was applied on both Se cathode and Li metal anode by adding FEC additive into an ether-based electrolyte (Figure 1.15a) [153]. The solid protective layer can reduce the dissolution of Se cathode materials as well as guide uniform decomposition of Li_2Se (Figure 1.15b,c). To combine the merits of both liquid and SSEs, gradient solidification of conventional ether-based liquid electrolytes was designed by Guo's group, through a facile in situ interfacial polymerization (IsIP) process (Figure 1.15d,e) [154]. As a result, a polymer-based gel electrolyte formed on the surface of the cathode, while the liquid electrolyte was preserved inside the cathode. The hybrid electrolyte configuration enables moderate dissolution of discharge products inside the cathode while preventing the shuttle of soluble intermediates away from the cathode. As a result, Li–Se batteries deliver a

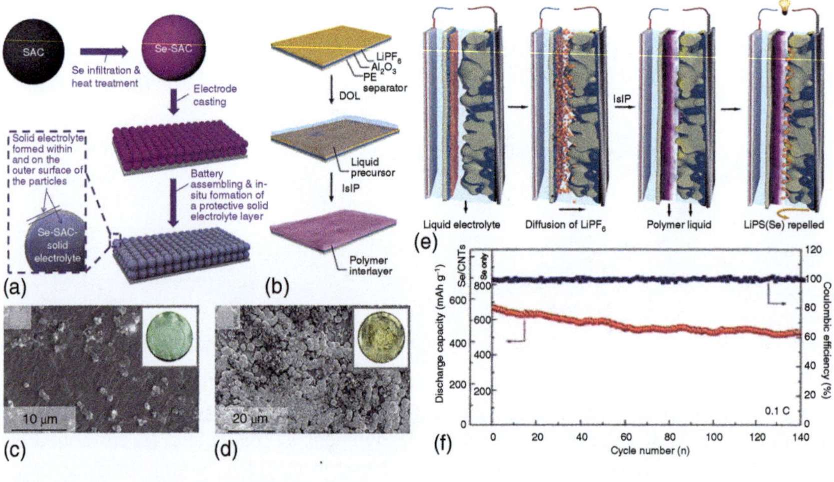

Figure 1.15 (a) Schematic illustration of formation of the protective layer on the Se cathode; SEM images of Li metal after one cycle with (b) FEC-enriched electrolyte and (c) a normal electrolyte. Source: Lee et al. [153] / with permission of Royal Society of Chemistry. (d) Schematic illustration showing in situ interfacial polymerization of a liquid electrolyte realized with the $LiPF_6/Al_2O_3$/PE separator. Source: Lee et al. [153] / with permission of Royal Society of Chemistry.; (e) schematic illustration showing the IsIP process of DOL induced by the $LiPF_6/Al_2O_3$/PE separator; (f) cycling performance and coulombic efficiency of Li–Se batteries with IsIP electrolyte at 0.1 C. Source: Reproduced with permission Wang et al. [154].

high initial specific capacity of 670 mAh g^{-1} and a high capacity retention of 83.5% after 140 cycles (Figure 1.15f).

Selenium has gained interest as cathode for Li–Se batteries due to its capacity density and high electronic conductivity. Numerous efforts have been made to designing nanostructured Se composite cathodes, exploring suitable electrolyte systems to suppress the shuttle effect, and improving the cycling stability. However, to promote the application of Li–Se batteries needs a lot of effort. For example, the mechanism of Se cathode in carbonate electrolytes is not clear. Some works indicate a single-conversion process, while others indicate a multiple-reaction process. In addition, due to the insolubility of polyselenides in carbonate electrolytes, in-depth researches need to be done with carbonate electrolytes instead of ether-based electrolytes. Finally, anode safety is also important in practical applications. More works on anode protection need to be designed, such as electrolyte additives, SSEs, or artificial SEI layer.

1.2.3.2 Lithium–Tellurium Batteries

Similar to Se, Te cathodes also receive relatively more attention [120, 155–158]. The huge problems are the "shuttle effect" of polytellurides and large volume expansion during cycling processes. To overcome these problems, the common strategies are to confine active materials into porous host materials to trap polytellurides and to accommodate the volume change [159].

He's group designed a rGO wrapped ultrathin tellurium nanowires (TeNWs) via a facile hydrothermal method as Te cathode for Li–Te batteries (Figure 1.16a) [160]. Due to the 3D framework and high conductivity of rGO, the 3DGT aerogel offers highly efficient channels for electron transfer and ionic diffusion, as well as good volume ratio for volume expansion (Figure 1.16b,c). Thus, the 3DGT shows a high initial capacity of 1685 mAh cm^{-3} at 1 C after 500 cycles (Figure 1.16d), showing remarkable long-cycle performance at high current density. Despite the nanostructured host materials, the cathode structure is also important to improve electrochemical performance. For example, some researchers confined Te into pores of host materials [164, 165], while others coated Te on the surface of host materials [166]. To compare these two different cathode structures, carbon nanotube aerogel was prepared by Yu and coworkers to host Te as cathode for lithium batteries [161]. Two different configurations of cathodes are designed: in configuration I, direct dropping Te solution on the surface of carbon nanotube aerogel; in configuration II, penetrating Te into the inner pores of carbon nanotubes (Figure 1.16e). When used as cathode materials for Li–Te batteries, configuration II exhibits much improved cycling performance due to the space confinement of nanopores to trap active Te molecules and the discharge product, while the elastic nature of the cathode helps to alleviate the deformation stress upon cycling.

Although the Te/carbon materials have exhibited excellent cycling stability, theoretical gravimetric capacity of Te is relatively low (420 mAh g^{-1}) [167, 168]. To address this problem, Te is combined with S element to form Te_xS_y composites, which can offer the merits of both elements, such as high conductivity of Te and high theoretical capacity of S [169]. Zhou's group confined heteroatomic Te_xS_{1-x}

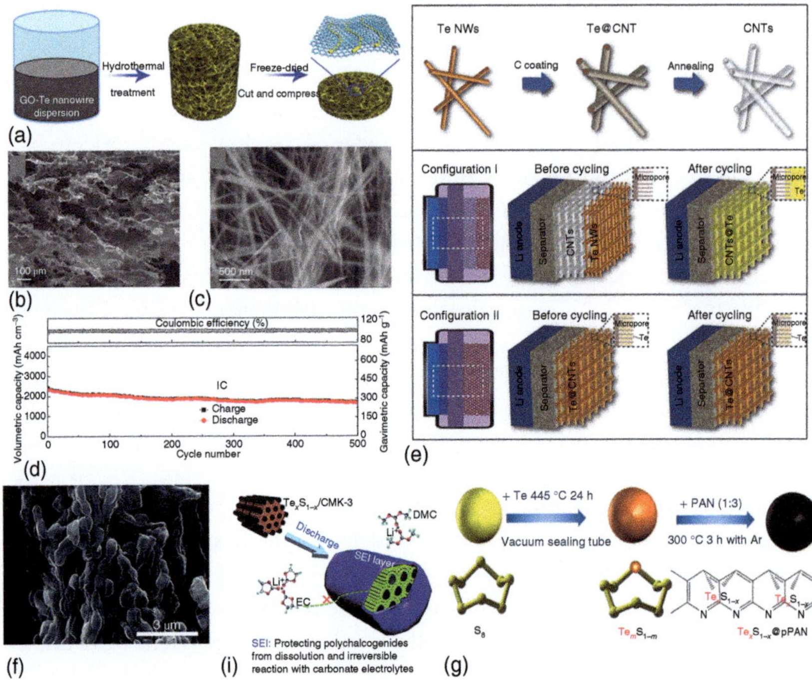

Figure 1.16 (a) Illustration of the formation process of the 3DGT aerogel and schematics of fabricating a self-supporting electrode; (b) low- and (c) high-magnification SEM images of 3DGT. Source: He et al. [160] / with permission of American Chemical Society. (d) Schematic illustration showing the typical synthesis procedures of the CNTs. (e) Configuration I using the TeNW/CNT cathode and its working principle. (f) Configuration II using the Te@CNT cathode and its working principle. Source: Xu et al. [161] / with permission of John Wiley and Sons. (g) SEM images of $Te_{0.1}S_{0.9}$/CMK-3; (h) schematic diagram of the Te-induced formation of SEI layers on the Te_xS_{1-x}/CMK-3 electrode surfaces. Source: Reproduced with permission Sun et al. [162]. (i) Synthetic schematic and the morphology characterization of $Te_{0.04}S_{0.96}$@pPAN. Source: Reproduced with permission Li et al. [163].

molecules into ordered mesoporous carbon CMK-3 via a facile melt-impregnation route (Figure 1.16g) [162]. The Te—S bonds in the heteroatomic Te_xS_{1-x} molecules offer higher intrinsic electrical conductivity and electrochemical reaction activity than the homoatomic S_8 molecules. Moreover, Te-containing polychalcogenide intermediates could help to form an SEI layer on Te_xS_{1-x}/CMK-3 surfaces, which can prevent polychalcogenides from the shuttle effect and side reactions with the carbonate solvent (Figure 1.16h). In addition, the mesopores on CMK-3 could confine the polychalcogenide intermediates. As a result, the $Te_{0.1}S_{0.9}$/CMK-3 composites can maintain a high reversible capacity of 845 mAh g^{-1} after 100 cycles at 250 mA g^{-1}.

To improve the slow kinetics of sulfurized polymer cathode [170, 171], Xie and coworkers used Te as an eutectic accelerator in S@pPAN to accelerate the redox conversion and prevent polysulfides dissolution (Figure 1.16i) [163]. In the Te-doped S@pPAN cathodes, Te can be uniformly distributed through Te—S bonds.

In addition, Te could contribute to improve capacity during the cycling process. Thus, Te-doped S@pPAN cathodes delivered high capacities and stable cycling over 600 cycles in the ether electrolyte (0.05% decay per cycle).

Until now, Te cathodes have received less attention than S cathodes due to their low theoretical gravimetric capacity. Heteroatomic Te_xS_{1-x} composites may be a promising candidate to improve gravimetric capacity. Although great achievements have been made on Te cathodes, the cycling stability is still far away from practical applications. Therefore, more efficient host materials and optimized systems need to be largely explored.

1.2.4 Lithium–Iodine/Bromine Batteries

With the emergence of hybrid grids and electric vehicles, there is an urgent need to develop energy storage devices with high energy density and power density. Nowadays, Li metal batteries paired with conversion type cathodes, such as S and O_2, have drawn much attention. But undeveloped discharge capacity and poor electrochemical reversibility of S and O_2 cathodes make them hard for practical applications. The shuttle effect of intermediates in Li–S batteries and the slow kinetic reactions of O_2 cathode need to be addressed before their commercialization. Therefore, exploring advanced novel batteries with redox reaction is urgently needed. Halogens (I_2 and Br_2, but not F_2 and Cl_2, which are gases and toxic) can be used in energy storage systems [172]. Bromine and iodine can provide relatively high specific capacities of 310 and 211 mAh g^{-1}, respectively. Moreover, the redox pairs of Br^{3-}/Br^- and I^{3-}/I^- have good reversibility with redox potentials at 1.07 and 0.54 V vs. SHE (standard hydrogen electrode). Although the specific capacity is low, but high voltage output can lead to high energy density. In addition, I_2 and Br_2 with good chemical/physical stability and low cost seem to be the promising alternative cathodes. In this section, we summarize the mechanism of Li–I_2/Br_2 batteries, then discuss the strategies made to improve electrochemical performance, and finally provide some suggestions.

1.2.4.1 Lithium–Iodine Batteries

1.2.4.1.1 Mechanism of Li–I_2 Batteries

Actually, as an abundant source in oceans, iodine has a long and profound history as a cathode material applied in batteries [173, 174]. The fundamental redox reactions involving iodine-based batteries can be described as follows (Figure 1.17) [175]:

On discharge process:

$$3I_2 + 2e^- = 2I_3^-$$

$$I_3^- + 2e^- = 3I^-$$

On charge process:

$$3I^- - 2e^- = I_3^-$$

$$2I_3^- - 2e^- = 3I_2$$

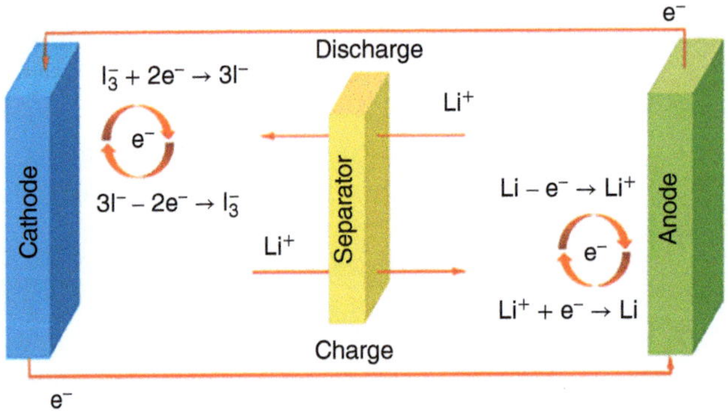

Figure 1.17 The working concept of the I_3^-/I^- redox reaction in an aqueous Li-I_2 battery. Source: Reproduced with permission Ma et al. [175].

The formation of soluble I_3^- ions and the dissolution of I_2 in the electrolyte decrease the utilization of active materials and lead to the shuttle effect, which can cause the corrosion of Li metal anodes, resulting in low columbic efficiency, poor cycling performance, and short lifetime of Li–I_2 batteries, which are the bottlenecks of Li–I_2 batteries [13, 173, 176].

1.2.4.1.2 Cathode and Interlayer Modification of Li–I_2 Batteries

To solve the aforementioned problems, numerous researches have been done. Similar to S cathodes, carbonaceous materials are widely used as matrixes to host I_2 [177–179]. In 2017, Liu's group applied active graphene (AG) to load high mass of I_2 (56 wt%), through an in situ deposition method (Figure 1.18a) [180]. The AG has a large specific surface area, abundant micropores and mesopores, free interparticle voids, and unimpeded ion diffusion channels. The as-prepared high mass loading I_2-AG composite delivers high capacity (218 mAh g^{-1} at a 1 C), stable cycling stability (capacity fading rate of only 0.023% per cycle), and good rate capability (184 mAh g^{-1} at a 20 C). These results indicate the nanopores in AG could effectively protect the I_2 from dissolution and improve the cycling stability. Recently, they further coated the I_2/AG with several kinds of polymers, such as methyl-beta-cyclodextrin (MβCD), polyvinylpyrrolidone (PVP), and amylose corn starch (ACS) [182]. The polymer can chemically react with I_2 to form inclusion complexes on the surface of I_2/AG composites. The formed inclusion complexes can effectively inhibit the dissolution of iodine into the electrolyte. As a result, the polymer-modified I_2/AG composites show enhanced electrochemical performance in terms of cycling stability and coulombic efficiency. Adding a block interlayer between cathode and separator is a direct way to suppress the shuttle effect [183]. Zhang's group synthesized honeycomb-like carbon materials (HCMs) via a "salty" thermal treatment of coffee extract to be applied as an interlayer to block the dissolution of I_2 (Figure 1.18b) [181]. The prepared HCMs had well-developed and interconnected ultrathin 3D frameworks with uniform distribution of nanopores (Figure 1.18c,d). This structure provides high surface area to physically block iodine species and fast ion transportation channels. As a result,

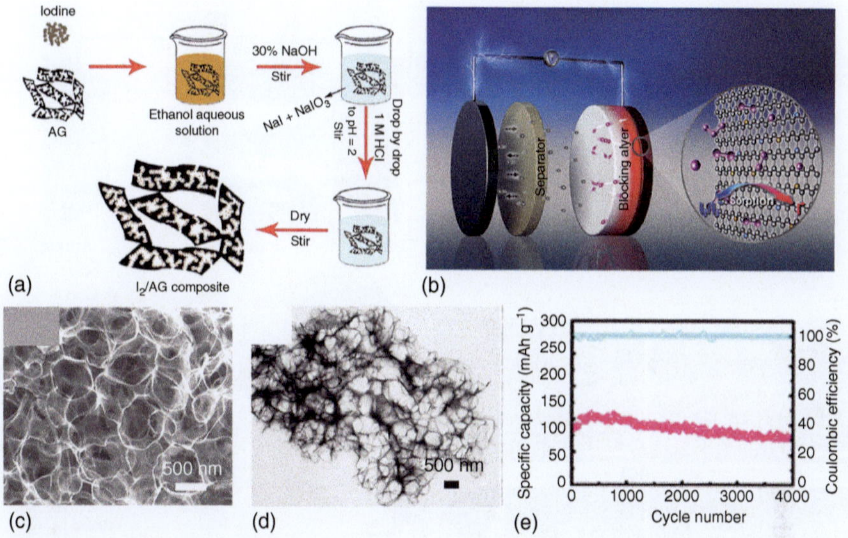

Figure 1.18 (a) Schematic depicting the in situ deposition route for fabricating the I_2-AG composites. Source: Reproduced with permission Zhang et al. [180]. (b) illustration of the working mechanism of the blocking layer in a rechargeable Li-I_2 battery; (c) SEM and (d) TEM images of synthesized HCM-5; (e) cycling performance of the as-constructed Li-I_2 batteries with HCM-5 blocking layer at a rate of 50 C. Source: Su et al. [181] / with permission of John Wiley and Sons.

the as-prepared Li–I_2 battery delivered a high capacity of 120.2 mAh g^{-1} after 4000 cycles even at a high 50 C rate (Figure 1.18e).

It is known that the reaction between nonpolar carbon and polar iodine species is poor, which is not enough to adsorb them. Thus, introducing polar functional groups into carbon matrixes is a promising way to further improve cycling stability of Li–I_2 batteries, such as heteroatom doping to increase surface polarity. Zhang's group prepared a self-standing nitrogen, phosphorus codoped carbon cloth used as a host for LiI material (Figure 1.19a) [184]. Heteroatom doping can provide extra active sites for strong adsorption of iodine species and the carbon scaffolds have abundant hierarchical pores leading to high surface area and contributing to capacity through capacitive effect (Figure 1.19b,c). These merits bring high flexibility and reversibility to Li–I_2 batteries, delivering a high capacity of 386 mAh g^{-1} while retaining 84.5% capacity after 2000 cycles. Apart from carbonaceous materials, metal-based materials were used to chemically adsorb soluble iodine species to improve the cycling stability of metal-I_2 batteries [186]. Lai's group applied a $Ti_3C_2T_x$ MXene foam as a highly effective interface layer for Li–I_2 batteries (Figure 1.19d) [185]. SEM images indicate that the prepared $Ti_3C_2T_x$ MXene exhibits a loose structure comprising ultrathin sheets, providing rich active sites to adsorb iodine species (Figure 1.19e). The thickness of $Ti_3C_2T_x$ MXene foam is about 120 μm, which is thick enough to block the transportation of soluble iodine species (Figure 1.19f). In addition, theoretical calculations revealed that the abundant functional groups on the surface of $Ti_3C_2T_x$ MXene sheets provide strong chemical

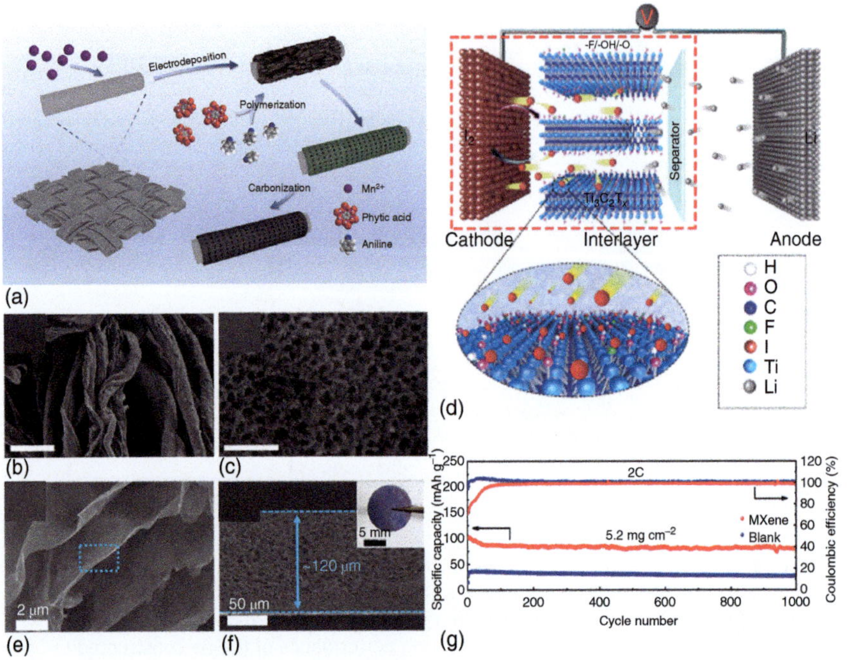

Figure 1.19 (a) Schematic illustration for the preparation process of CC-NP; (b) and (c) SEM images of CC-NP. Source: Li et al. [184]. Reproduced with permission of (2019) Springer. (d) Schematic of the reaction mechanisms of a Li-I_2 battery with a MXene interface layer; (e) SEM image of $Ti_3C_2T_x$ MXene sheets; (f) the cross-sectional SEM images of $Ti_3C_2T_x$ MXene interface layer with an inset showing a digital photo of the $Ti_3C_2T_x$ MXene interface layer; (g) long-term cycling performance and the corresponding coulombic efficiency of I_2 cathode at a mass loading of 5.2 mg cm^{-2}. Source: Sun et al. [185]. Reproduced with permission of (2020) American Chemical Society.

binding to iodine species, suppressing the shuttle effect. Moreover, high electronic conductivity of $Ti_3C_2T_x$ MXene sheets facilitates fast iodine redox reactions. As a result, the Li–I_2 battery with I_2@N-doped graphene composite cathode containing 70% I_2 and $Ti_3C_2T_x$ MXene foam as interlayer can cycle stably for over 1000 cycles at a rate of 2 C, even at an ultrahigh loading mass of 5.2 mg cm^{-2} (Figure 1.19g).

1.2.4.1.3 Anode Protection of Li–I_2 Batteries

Lithium dendrites caused by unstable SEI and the nonuniform deposition of Li ions on Li metal anodes is another challenge for practical applications of Li–I_2 batteries [187, 188]. To address this issue, Zhang's group synthesized an N, P codoped carbon (NPCC) skeleton by interfacial polymerization of polyaniline on the carbon cloth (Figure 1.20a) [189]. Then, molten lithium was loaded on the surface of 3D modified carbon cloth. The surface of prepared NPCC showed interconnected carbon nanorod networks (Figure 1.20b), indicating the successful modification of heteroatoms. In addition, the N and P elements were uniformly distributed on the surface of NPCC (Figure 1.20c), leading to a lithiophilic surface to uniformly load Li

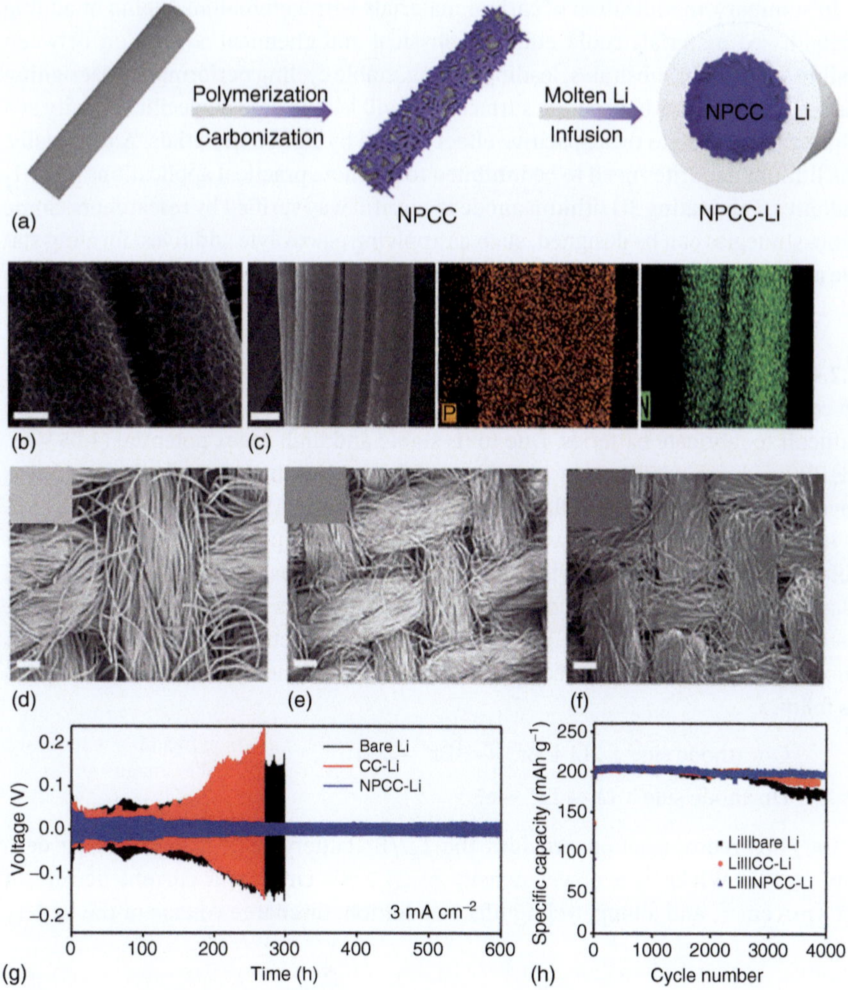

Figure 1.20 (a) Schematic illustration of the synthesis process of NPCC-Li; (b) SEM images of NPCC; (c) uniform distribution of N and P on NPCC; SEM images of NPCC with different Li loading: (d) 3.5, (e) 7.5, and (f) 12 mg; (g) galvanostatic charge–discharge curves of symmetric cells assembled with Li, CC-Li, and NPCC-Li electrodes at current density 3 mA cm^{-2}; (h) cycling performances of full cells with iodine cathode. Source: Li et al. [189] / with permission of John Wiley and Sons.

metal. Due to the uniform, lithiophilic surface, the Li metal can be uniformly coated on the carbon skeleton, even with a high mass loading of 12 mg cm^{-2} without any agglomeration (Figure 1.20d–f). Moreover, the 3D porous structure formed would lower the local current density for reversible deposition and stripping of Li metal. The Li symmetric cells could stably cycle for over 600 hours at a current density of 3 mA cm^{-2} (Figure 1.20g). When coupled with iodine cathode, the assembled battery showed an impressive long-term cycling stability with a capacity retention of 99% after 4000 cycles (Figure 1.20h).

In summary, modification of carbon materials with heteroatom doping or adding metal-based materials could enhance physical and chemical adsorption between iodine and carbon substrates, leading to long, stable cycling performance. Designing carbon materials with ultrahigh surface area will lead to higher specific capacity at a high scan rate due to the capacitive effect offered by carbon materials. Additionally, the lithium dendrites need to be inhibited to promote practical applications of Li–I_2 batteries. Fabricating 3D lithium anode is a useful way verified by researchers. Some more strategies can be designed, such as applying electrolyte additives, forming stable artificial SEI layer. If all these problems can be solved, the commercialization of Li–I_2 batteries with high energy density is not far.

1.2.4.2 Lithium–Bromine Battery

In comparison with iodine, bromine as a liquid at room temperature makes it more difficult to fabricate batteries. Due to its stable and high redox potential (1.05 V vs. SHE), good reversibility, fast kinetics, and high solubility in water, bromine was investigated in redox flow cell, such as Zn–Br batteries [190].

In 2014, a Li//Br battery was designed by Yu's group [191]. The aqueous LiBr solution was used as electrolyte, with liquid Br_2 or Br^{3-} as the positive electrode and glassy carbon as the current collector. Lithium metal was coated with a home-made gel polymer electrolyte (GPE) and a ceramic solid electrolyte (LISICON) film, as shown in Figure 1.21a. The redox reactions of the designed Li//Br battery are given as follows:

$$\text{Oncathode side}: 2\text{Li} + \text{Br}_3^- \leftrightarrow 3\text{Br}^- + 2\text{Li}^+$$

$$\text{On anode side}: \text{Li} \leftrightarrow \text{Li}^+ + \text{e}^-$$

Br^-/Br_3^- redox reaction provides the Li//Br battery with a high energy density of 1220 Wh kg^{-1}, a power density of 29.7 mW cm^{-2} at a current density of 12.8 mA cm^{-2}, and a long cycling life. In addition, discharge voltage of the battery

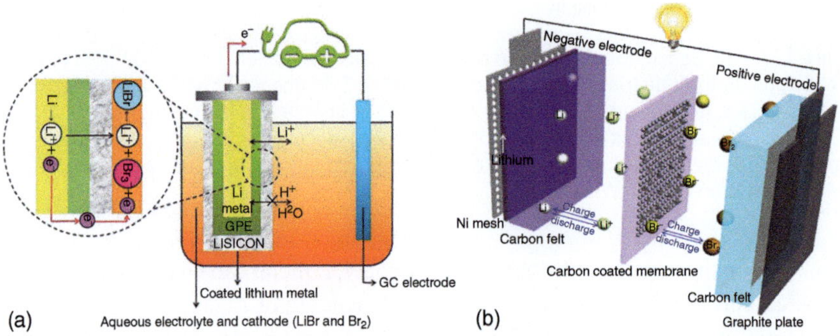

Figure 1.21 (a) Schematic illustration of our rechargeable Li//Br battery composed of Li metal coated with GPE and LISICON as the negative electrode, aqueous LiBr solution (1 M) as the electrolyte, and 0.1 M liquid Br_2 or Br_3^- as the positive electrode with glassy carbon as the current collector. Source: Reproduced with permission Chang et al. [191]. (b) Schematic illustration of a nonaqueous Li/Br battery. Source: Reproduced with permission Xi et al. [192].

is high (3.96 V) and very stable, without degradation during the discharge process. These advantages enable Li/Br batteries competitive to commercial Li-ion batteries. Recently, a nonaqueous lithium bromine rechargeable battery is proposed by Zhang's group [192]. On the Li metal anode side, Ni mesh was used as current collector and carbon felt was used to carry the electrolyte. On the cathode side, graphite plate was applied as the current collector and carbon felt was used to adsorb the electrolyte. The electrolyte consisted of 2 M LiBr in 1.3 M LiTFSI DOL/DME solution. A carbon-coated membrane was applied as a separator (Figure 1.21b). The designed Li–Br batteries operated via the redox conversion of Br_2/Br^- and Li^+/Li pairs, as follows:

$$\text{Cathode side}: Br_2 + 2e^- \leftrightarrow 2Br^-$$

$$\text{Anode side}: Li - e^- \leftrightarrow Li^+$$

Fast redox kinetics and good stability of Br_2/Br^- offer the Li/Br battery with high output voltage (~3.1 V), high electrolyte concentration (3.0 M), and practical energy density (232.6 Wh kg^{-1}). The nonaqueous Li/Br battery can be readily scaled up as alternatives of Li-ion batteries.

Although great achievements have been made on Li–Br batteries, compared with Li–I_2 batteries, the research on Li–Br batteries are limited due to the liquid state of Br_2 at room temperature. To find stable supporters or host materials for Br_2 at room temperature is a big challenge with Li–Br_2 batteries. Although high energy density is achieved, specific capacity of Li–Br batteries is ultralow. Adding redox pairs, such as Cl/Cl^- with a theoretical capacity up to 755 mAh g^{-1} can further improve the capacity as well as energy density. Another challenge to be solved is the shuttle effect, which has not been reported in Li–Br batteries. Applying matrixes with both physical and chemical adsorption functions is a promising way. As for Li metal anodes, Li dendrites should be considered to improve the long-term stability of Li–Br batteries. The strategies already applied in other Li metal batteries can be used as reference, such as adding electrolyte additives, forming artificial SEI layer, or synthesizing 3D matrixes to load Li metal. These are alternative ways to develop advanced rechargeable Li–Br batteries with high energy density and stable reversibility.

1.2.5 TMO Batteries

The Li-ion batteries have been dominating the green energy markets for several decades, which consist of graphite as anode and lithium-containing transition metal compounds with different structures, such as olivine ($LiMPO_4$:M = transition metal), layered ($LiMO_2$), or spinel (LiM_2O_4) materials [193]. However, the state-of-the-art commercial Li-ion batteries have already reached their theoretical capacity limitation. The emergence of grids and electric vehicles drives the researchers to find electrode materials with higher capacity and energy density [194]. Li metal is thus believed to be the ultimate choice to replace graphite anodes due to its high theoretical specific capacity (3860 mAh g^{-1}) and lightweight (0.53 g cm^{-3}). For cathode materials, compared with lithium-containing transition metal oxides, lithium-free transition metal oxides (TMOs) are also considered

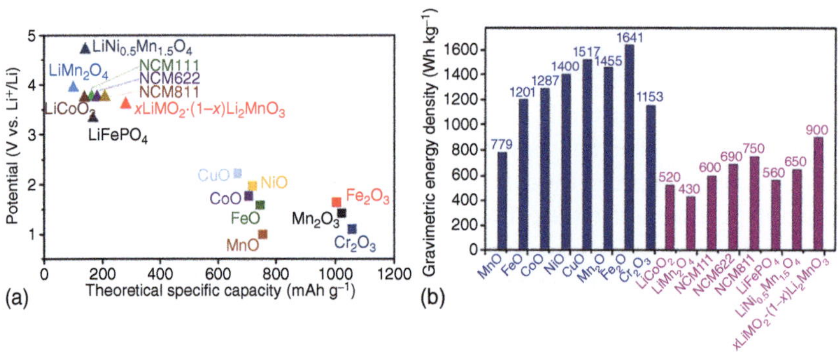

Figure 1.22 Fundamentals of conversion reaction-type TMOs. (a) Theoretical specific capacity and voltage and (b) energy density for the TMOs compared to conventional cathodes. Source: Reproduced with permission Wu et al. [196].

promising candidates [195, 196]. As shown in Figure 1.22, the TMO materials offer much higher specific capacity than that of conventional intercalation/phase transition-type cathodes such as LFP, $LiCoO_2$ (LCO), NCM, and even Li-rich Mn-based oxides. In addition, when coupled with Li metal, TMOs have the potential to offer energy density of 1200–1700 Wh kg^{-1}, which is two or three times that of commercialized cathode materials of Li–ion batteries. Although the operating voltage of TMOs is relatively lower, their high specific capacity still can yield higher energy density.

Several TMOs have been prepared as cathode materials for Li metal batteries. Deng's group prepared three kinds of conversion reaction-type TMOs (Fe_2O_3, Mn_2O_3 or CuO) as Li-free cathodes for solid-state Li metal batteries, in which PEO, lithium garnet $Li_{6.4}La_3Zr_{1.4}Ta_{0.6}O_{12}$ (LLZTO), or $Li_{1.5}Al_{0.5}Ge_{1.5}(PO_3)_4$ (LAGP) are employed as SSEs [196]. The cathodes offer good interfacial compatibility with the SSEs, leading to excellent electrochemical performance of the TMOs/SSEs/Li solid-state cells. The prepared Fe_2O_3 has a small size with nanostructures (Figure 1.23a); when used as cathode in the solid-state Li metal battery, the Fe_2O_3/PEO/Li cell delivers a high discharge capacity of 990 mAh g^{-1} at 10 mA g^{-1} at 60 °C, which is comparable to the Fe_2O_3 liquid cell. When cycling at 200 mA g^{-1} a stable capacity of 733 mAh g^{-1} is achieved after 1100 cycles for the Fe_2O_3/PEO/Li cell, while the Fe_2O_3 liquid cell sustains only 860 cycles until failure (Figure 1.23b). The MnO_2 sheets showing a high specific capacity of 616 mAh g^{-1} is regarded as a promising cathode for Li metal batteries. But the structural collapse during cycling hinders its further application. To improve the structural stability of MnO_2 nanosheets, Deng and coworkers designed a three-dimensional porous NiO nanorod pillared δ-MnO_2 nanosheets integrated nanostructure as cathode materials for Li metal batteries (Figure 1.23c) [197]. The NiO pillars not only can prevent restacking of δ-MnO_2 nanosheets but also provide more electrochemically active sites and thermodynamically favorable insertion pathways for the lithium ions. Thus, the as-synthesized NiO pillared δ-MnO_2 nanosheets delivered a relatively high specific capacity of 185 mAh g^{-1} and maintained a capacity of 116 mAh g^{-1} after 200 cycles at 100 mA g^{-1} (Figure 1.23d).

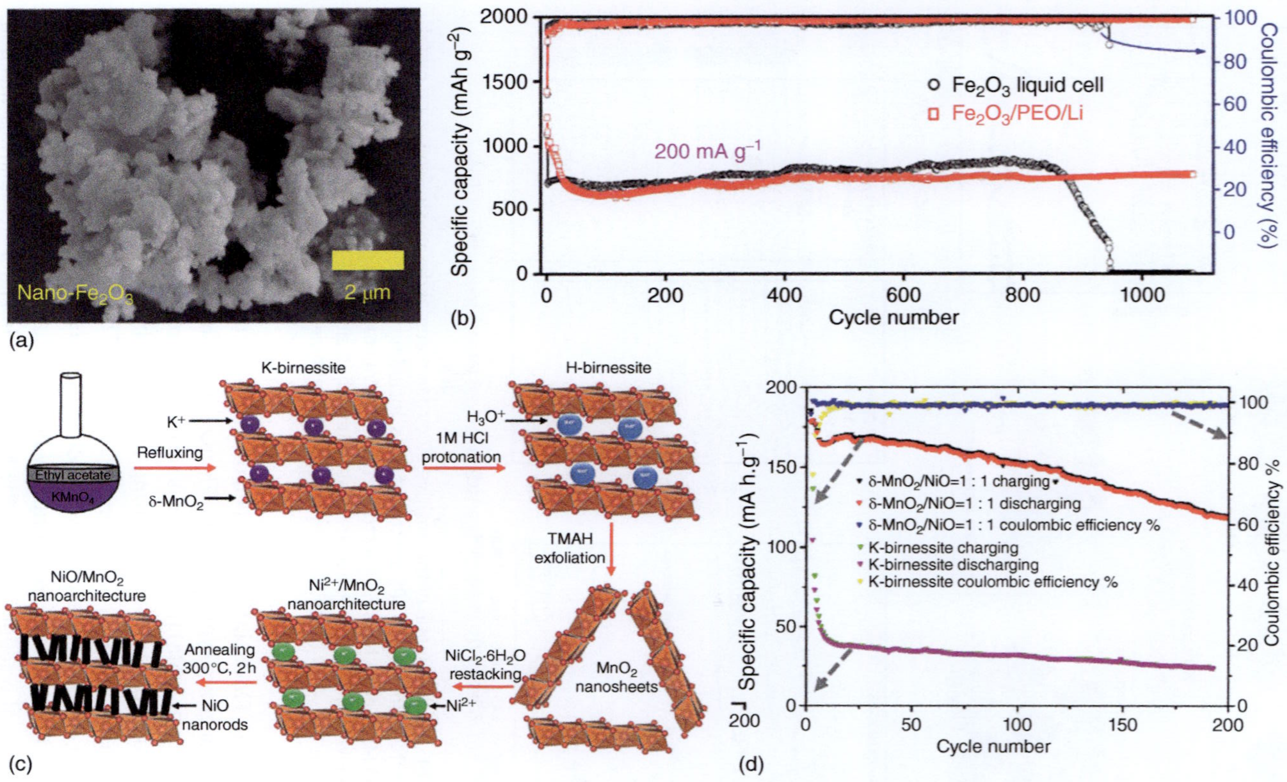

Figure 1.23 (a) SEM image of prepared nano-Fe_2O_3 and (b) cycling performance of Fe_2O_3/PEO/Li and Fe_2O_3 liquid cells for 1100 cycles at 200 mA g^{-1}. Source: Wu et al. [196] / with permission from Elsevier. (c) Schematic showing the steps and structural changes involved in the synthesis of 3-D nanoarchitectures of NiO nanorod pillared δ-MnO_2 nanosheets; (d) cycling performance of δ-MnO_2 nanosheets and NiO pillared δ-MnO_2 nanosheets as lithium-free cathode materials at 100 mA g^{-1}. Source: Reproduced with permission Hewavitharana et al. [197].

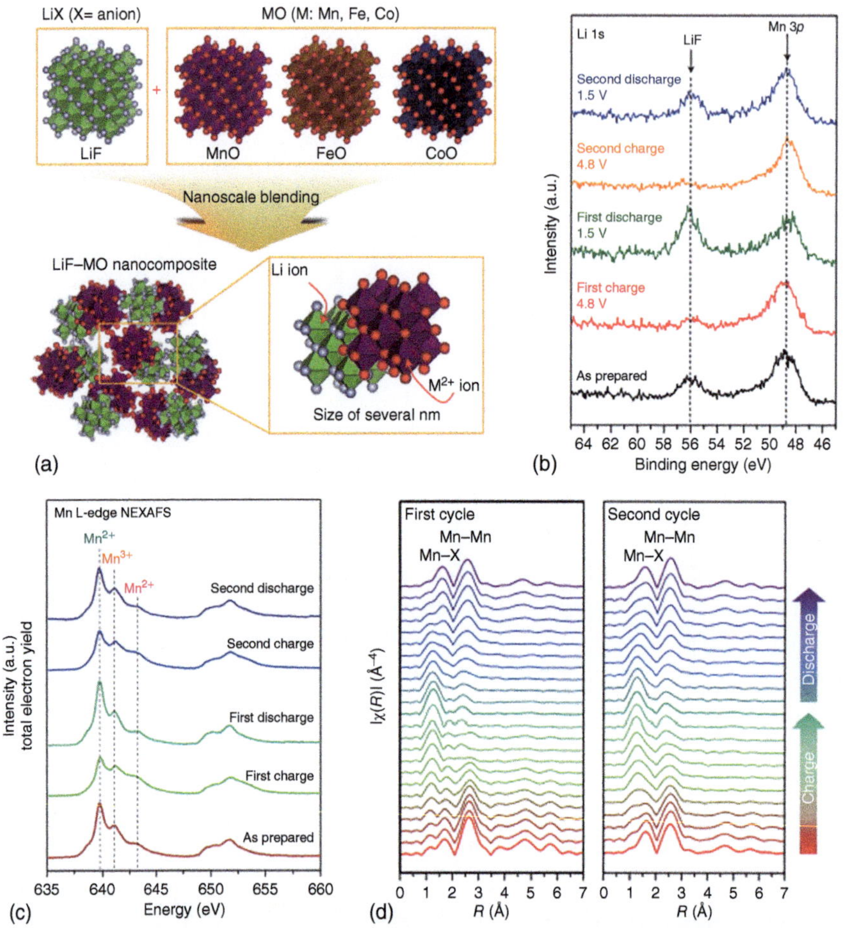

Figure 1.24 (a) Schematic of the positive electrode material design strategy using metal monoxide (MO); (b) ex situ XPS spectra of Li 1s, including the range of Mn 3p binding energy; (c) Mn L-edge NEXAFS spectra of the LiF–MnO nanocomposite from total electron yield mode with different electrochemical states; (d) in situ EXAFS spectra during two cycles in the 1.5–4.8 V range. Source: Reproduced with permission Jung et al. [198].

Except pure TMOs, LiF was applied on TMOs to improve the operating voltage. In 2017, Kang's group synthesized LiF/MnO composites via high-energy ball-milling method (Figure 1.24a) [198]. There is no chemical reaction between MnO and LiF. To understand the role of LiF during discharge/charge process, the Li 1s region was analyzed by XPS, as shown in Figure 1.24b. The signal of LiF at 56.0 eV disappeared at charge and recovered at discharge, indicating the F ion can be bound to the Mn ion to satisfy charge neutrality when oxidation occurs. To further reveal the electrochemical reaction between the MnO and F ions, further analysis was done by soft X-ray absorption spectroscopy (Figure 1.24c). On charging, the Mn^{3+} and Mn^{4+} signals increased at the expense of the Mn^{2+} signal. The reversible change was observed during the subsequent electrochemical cycling, which indicates that not only Mn^{2+}/Mn^{3+} but also Mn^{3+}/Mn^{4+} redox reactions are involved in the electrochemical reaction. To observe the local structure evolution, extended X-ray absorption fine structure (EXAFS) analysis was used, indicating that the structural transition is reversible during electrochemical cycling. Therefore, the reaction mechanism of the LiF/MnO nanocomposite can be called a "surface conversion reaction," which is described as follows:

$$LiF + Mn^{2+}O \rightarrow Li^{+} + e^{-} + Mn^{3+}O\text{–}F$$

LiF functions not only as a Li source for cathode materials but also as an anion source for charge compensation when oxidation and reduction of a transition metal ion occur. LiF can be regarded as a stabilizer of the oxidized Mn^{3+} (or Mn^{4+}) ion by providing F^{-} ions from the MnO perspective. In addition, the MnO can be regarded as an assistant that promotes the decomposition of LiF. As a result, the LiF/MnO composites exhibit an average voltage of over 3 V (vs. Li^{+}/Li).

Although some TMOs have been studied as cathode materials for Li metal batteries, compared with metal sulfides or fluorides, the development of TMOs is still in its infant stage. There are some problems that need to be addressed to promote the application of TMOs. First, the cycling stability of TMOs in liquid Li metal batteries is inferior. The mechanism and strategies should be developed, such as applying in situ analysis techniques and surface coating. The rate capability of TMOs in solid-state Li metal batteries should be improved by synthesizing SSEs with high ionic conductivity.

1.3 Introductive Electrolytes

Li metal has been intensively investigated as anode due to its ultrahigh theoretical specific capacity (3860 mAh g^{-1}) and the most negative electrochemical potential (~3.040 V vs. SHE). When paired with cathode materials, Li metal batteries can deliver higher capacity and higher operating voltage, leading to higher energy density, which make lithium metal batteries (LMBs) promising power devices for long-term vehicles or large-scale grids to replace commercial Li-ion batteries [199]. Despite of these advantages, several problems are also found that hinder practical applications of LMBs [200]. As shown in Figure 1.25a, the Li metal can react with the electrolyte to form an SEI layer on the surface of the Li metal anode. During plating process, volume expansion will occur, which can crack the unstable SEI layer. The new exposed Li metal will form dendrites. In the subsequent striping process, the dendrite-like Li metal will be isolated to form "dead" Li. During the long-term cycling process, a porous Li metal with thick SEI and accumulated dead Li will be obtained, which can lead to low columbic efficiency, capacity fading, and failure of cells. To overcome these concerns, various approaches have been reported in the literature to make Li metal anodes with high stability, including electrolyte formulation [202], SSEs [203], artificial stable SEI [204], and anode structure design [205]. Among them, electrolytes play a vital role in obtaining a mechanical stable SEI layer and suppressing the growth of Li dendrites [201, 206, 207].

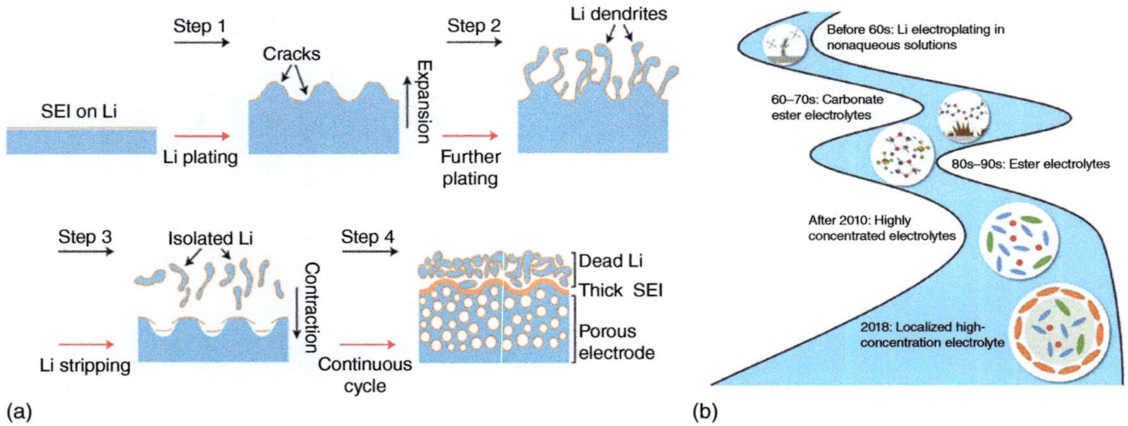

Figure 1.25 (a) Schematic showing the problems related to Li metal during the Li stripping/plating process. Source: Reproduced with permission Lin et al. [200]. (b) Schematic diagram for the development progress of liquid electrolytes in lithium metal batteries. Source: Reproduced with permission Jie et al. [201].

As shown in Figure 1.25b, the investigation of electrolyte for Li plating/striping can be dated back to the nineteenth century, using a nonaqueous electrolyte with LiCl and pyridine. Then, different salts and solvents were studied [208, 209]. After preliminary studies, aprotic solvents were proved to be useful to inhibit the reaction between Li metal and electrolytes. In the 1960s, due to the demand for high-energy-density storage systems, the electrolytes used in LMBs started to get attention. Electrolytes with alkyl carbonate esters such as propylene carbonate (PC) as solvents and $LiPF_6$ or $LiClO_4$ as Li salts were used in rechargeable batteries due to their good ionic conductivity [210]. However, carbonate ester-based electrolytes showed low coulombic efficiency, which hindered their further application [211].

In the late 1970s, ether electrolytes were reported, which could obtain high CE and stable reversibility of the Li electrode due to the formation of more suitable SEI layers with both organic and inorganic contents [212]. With the commercialization of Li–ion batteries with graphite as anode, the research on electrolytes for LMBs was limited. Recently, as graphite anodes are approaching their theoretical capacity, there has been a renaissance in LMBs. At this stage, excellent strategies have been applied to improve the cycling stability of both anode and cathode through electrolyte modifications. Adding suitable electrolyte additives is a simple and promising way to in situ form stable solid electrolyte interfaces on both anode and cathode, leading to enhanced cycling stability of LMBs [213]. For example, gradient SEI layer was formed by adding bisfluoroacetamide (BFA) into the electrolyte, consisting of a C–F rich surface and a LiF rich bottom layer (Figure 1.26a,b), which can contribute to the preferential reduction of BFA on Li metal surface due to higher reductive potentials (Figure 1.26c) [214]. As a result, the Li||Li batteries can stably cycle for 400 cycles at a current density of $1\,mA\,cm^{-1}$ (Figure 1.26d). The solvation structures of Li ions can also be changed by adding electrolyte additives, such as alkyl-triphenyl-phosphonium bromides (alkyl-TPPB) [215]. The presence of Br^- ions in the electrolyte can enlarge the reaction distance between Li ions with solvents (Figure 1.26e). Thus, the coordinate number of EC in solvation structures decreases, which can suppress the reduction of solvents and improve the stability of electrolytes. HCEs were also reported to improve the cycling stability of Li metal anodes [216, 217]. In addition, HCEs exhibit good oxidation stability, which can simultaneously improve the stability of cathode materials, such as $LiNi_{0.8}Mn_{0.1}Co_{0.1}O_2$ (NMC811). However, high cost and viscosity of HCEs make them unsuitable for practical applications. Therefore, diluents are added into electrolytes to form localized high-concentration electrolytes (LHCEs) [218]. LHCEs exhibit the advantages of HCEs, as well as decrease the cost and viscosity of HCEs, which are regarded as advanced liquid electrolytes for Li metal batteries.

Except these electrolyte modifications, some functional electrolytes were also studied, such as nonflammable electrolytes [219], high-voltage electrolytes [220], electrolytes operated at extreme conditions (such as high or low temperature) [221]. Significant advances from both the fundamental understanding and practical engineering have been made, but the reaction between Li metal and electrolyte is

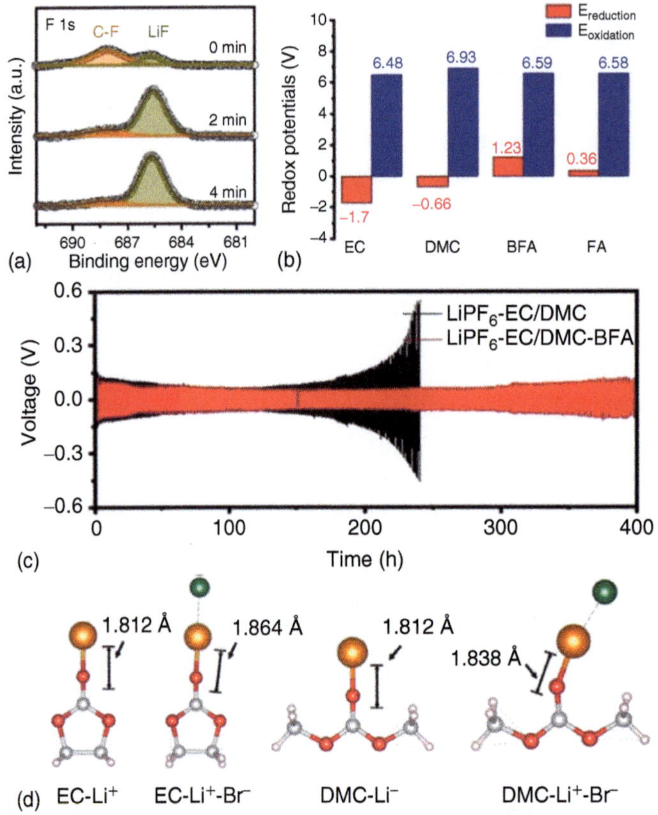

Figure 1.26 (a) XPS spectra of F 1s in Li||Li cells after 10 cycles in $LiPF_6$-EC/DMC-BFA electrolyte. (b) Theoretical calculation of redox potentials of EC, DMC, BFA, and FA molecules; (c) cycling performance of Li||Li cells using $LiPF_6$-EC/DMC-BFA electrolyte at a current density of 1 mA cm^{-2} with a cutoff capacity of 0.5 mAh cm^{-2}. Source: Reproduced with permission Li et al. [214]. (d) The comparison of the distance between Li^+ ion and O atom of the EC and DMC molecules with and without the Br^- ion coordinator. Source: Reproduced with permission Qi et al. [215].

not inhibited, which will affect the practical applications of LMBs with long-term requirements.

1.4 Prospects

Over the past several decades, LMBs have been regaining intensive research attention due to their high energy density compared with Li-ion batteries. Until now, practical applications of LMBs with different cathodes is not realized due to their inherent drawbacks. Therefore, to solve these detrimental issues and realize the practical deployment of LMBs, many problems need to be addressed:

(1) Mechanism: Clear understanding of the reaction mechanism is vital to further improve electrochemical performance of batteries. The mechanism of Li_2O_2

decomposition on a cathode catalyst in Li–O_2 batteries is not elucidated yet. Due to the sluggish kinetics of O_2, designing cathodes with high surface area as well as catalysts with high OER catalytic activity is important. In addition, for practical applications of Li–O_2 batteries, the effect of CO_2 in air should be considered, which can react with Li_2O_2 to produce Li_2CO_3. For Se cathodes, the redox reaction pathway in carbonate electrolytes is not clear. Some works indicate a single-conversion process, while others indicate a multiple-reaction process. In addition, due to the insolubility of polyselenides in carbonate electrolytes, in-depth research needs to be done on carbonate electrolytes rather ether-based electrolytes.

(2) Shuttle effect: The shuttle effect is a big problem in Li–S, Li–Se/Te, Li–I_2/Br_2 batteries. To solve this problem, designing cathode host materials with both physical and chemical adsorption properties is a promising way, such as heteroatom doping on carbonaceous materials and introducing metal-based polar species. Inserting a block interlayer is the simplest way. For practical applications, lean electrolytes and high mass loading should be considered, which are generally are paid less attention. Exploring SSEs with high ionic conductivity is the direct way to inhibit the shuttle effect.

(3) Li metal anodes: Due to random deposition of Li ions, the dendrite growth is the biggest challenge with Li metal batteries. More works on anode protection need to be done, such as electrolyte additives, SSEs, or artificial SEI layer on Li metal. To further understand anode and electrolyte degradation as well as the components and structures of electrolyte interfaces, new analytical tools and in situ techniques should be applied.

If these problems are solved, Li metal batteries will take great strides toward their practical applications.

References

1 Wu, X., Wang, J., Fei, D. et al. (2014). *Energy & Environmental Science* 7 (2): 513–537.

2 Lin, D., Liu, Y., and Yi, C. (2017). *Nature Nanotechnology* 12 (3): 194–206.

3 Whittingham, M.S. (2004). *Chemical Reviews* 35 (50): 4271–4301.

4 Dusastre, V. (2011). *Materials for Sustainable Energy*. Nature Publishing Group.

5 Hong, L., Wang, Z., Chen, L., and Huang, X. (2009). *Advanced Materials* 21 (45): 4593–4607.

6 Li, M., Lu, J., Chen, Z., and Amine, K. (2018). *Advanced Materials* 30 (33): *1800561*.

7 Daniel, I.C. and Besenhard, J.O. (2012). *Handbook of Battery Materials*, 2e. John Wiley & Sons.

8 Chen, L., Liu, Y., Ashuri, M. et al. (2014). *Journal of Materials Chemistry A* 2 (42): 18026–18032.

9 Reich, H.J., Bevan, M.J., Gudmundsson, B., and Puckett, C.L. (2002). *Angewandte Chemie International Edition* 41 (18): 3436–3439.

10 Sun, J., Du, Z., Liu, Y. et al. (2021). *Advanced Materials* 33 (10): 2003845.
11 Lu, J., Li, L., Park, J.B. et al. (2014). *Chemical Reviews* 114 (11): 5611–5640.
12 Lu, K., Hu, Z., Ma, J. et al. (2017). *Nature Communications* 8 (1): 1–10.
13 Xu, J., Ma, J., Fan, Q. et al. (2017). *Advanced Materials* 29 (28): 1606454.
14 Cheng, X.B., Zhang, R., Zhao, C.Z., and Zhang, Q. (2017). *Chemical Reviews* 117 (15): 10403–10473.
15 Shao, Y., Fei, D., Jie, X. et al. (2013). *Advanced Functional Materials* 23 (8): 987–1004.
16 Kwak, W., Rosy, D., Sharon, C. et al. (2020). *Chemical Reviews* 120 (14): 6626–6683.
17 Lyu, Z., Zhou, Y., Dai, W. et al. (2017). *Chemical Society Reviews* 46 (19): 6046–6072.
18 Yu, Z., Wang, H., Kong, X. et al. (2020). *Nature Energy* 5: 526–533.
19 Lim, H.K., Lim, H.D., Park, K.Y. et al. (2013). *Journal of the American Chemical Society* 135 (26): 9733–9742.
20 McCloskey, B.D., Bethune, D.S., Shelby, R.M. et al. (2011). *Journal of Physical Chemistry Letters* 2 (10): 1161–1166.
21 McCloskey, B.D., Speidel, A., Scheffler, R. et al. (2012). *Journal of Physical Chemistry Letters* 3 (8): 997–1001.
22 McCloskey, B.D., Scheffler, R., Speidel, A. et al. (2011). *Journal of the American Chemical Society* 133: 18038.
23 Chang, Z., Xu, J., and Zhang, X. (2017). *Advanced Energy Materials* 7 (23): 1700875.
24 Wang, D., Mu, X., He, P., and Zhou, H. (2019). *Materials Today* 26: 87–99.
25 Xiao, J., Mei, D., Li, X. et al. (2011). *Nano Letters* 11 (11): 5071–5078.
26 Jiao, W., Su, Q., Ge, J. et al. (2021). *Materials Research Bulletin* 133: 111020.
27 Huang, H., Luo, S., Liu, C. et al. (2017). *Journal of Alloys and Compounds* 726: 939–946.
28 Jung, J.-W., Choi, D.-W., Lee, C.K. et al. (2018). *Nano Energy* 46: 193–202.
29 Kim, M., Kim, D.W., Suk, J. et al. (2015). *Carbon* 93: 625–635.
30 Chang, Y., Dong, S., Ju, Y. et al. (2015). *Advanced Science* 2 (8): 1500092.
31 Cao, Y., Lu, H., Hong, Q. et al. (2019). *Carbon* 144: 280–288.
32 Nie, H., Xu, C., Zhou, W. et al. (2016). *ACS Applied Materials & Interfaces* 8 (3): 1937–1942.
33 Cao, D., Zhang, S., Yu, F. et al. (2019). *Batteries & Supercaps* 2 (5): 428–439.
34 Zhao, Y., Ding, L., Wang, X. et al. (2021). *Journal of Alloys and Compounds* 861: 157945.
35 Luo, N., Ji, G.-J., Wang, H.-F. et al. (2020). *ACS Nano* 14 (3): 3281–3289.
36 Shui, J.-L., Okasinski, J.S., Kenesei, P. et al. (2013). *Nature Communications* 4 (1): 1–7.
37 Zhang, X., Xie, Z., and Zhou, Z. (2019). *ChemElectroChem* 6 (7): 1969–1977.
38 Ko, J. and Yoon, Y.S. (2018). *Ceramics International* 45 (1): 30–49.
39 Yu, Y., Huang, G., Wang, J.Z. et al. (2020). *Advanced Materials* 32 (38): 2004157.
40 Adair, K.R., Zhao, C., Banis, M.N. et al. (2019). *Angewandte Chemie International Edition* 58 (44): 15797–15802.

41 Luo, Z., Zhu, G., Yin, L. et al. (2020). *ACS Applied Materials & Interfaces* 12 (24): 27316–27326.

42 Yu, Y., Yin, Y.-B., Ma, J.-L. et al. (2019). *Energy Storage Materials* 18: 382–388.

43 Liu, T., Leskes, M., Yu, W. et al. (2015). *Science* 350: 530–533.

44 Yang, L., Wang, X., Dong, S. et al. (2016). *Advanced Energy Materials* 6 (18): 1600751.

45 Yu, Y. and Zhang, X.-B. (2019). *Matter* 1 (4): 881–892.

46 Liu, B., Xu, W., Yan, P. et al. (2016). *Advanced Functional Materials* 26 (4): 605–613.

47 Wu, S., Zhu, K., Tang, J. et al. (2016). *Energy & Environmental Science* 9 (10): 3262–3271.

48 Qin, L., Zhai, D.Y., Lv, W. et al. (2017). *Nano Energy* 40: 258–263.

49 Manthiram, A., Fu, Y., Chung, S.-H. et al. (2014). *Chemical Reviews* 114 (23): 11751–11787.

50 Ruopian, F., Shiyong, Z., Zhenhua, S. et al. (2017). *Advanced Materials* 29 (48): *1606823.*

51 He, J. and Manthiram, A. (2020). *Advanced Energy Materials* 10 (3): 1903241.

52 Danuta, H.; Juliusz, U. (1962). Electric dry cells and storage batteries. U.S. Patent, No.3,043,896.

53 Rauh, R.; Pearson, G.; Brummer, S. (1977). *Rechargeability studies of ambient temperature lithium/sulfur batteries.* 12th Intersociety Energy Conversion Engineering Conference, American Nuclear Society, 1, 283–287.

54 Linden, D. and McDonald, B. (1980). *Journal of Power Sources* 5 (1): 35–55.

55 Yamin, H., Gorenshtein, A., Penciner, J. et al. (1988). *Journal of the Electrochemical Society* 135 (5): 1045.

56 Wang, J., Yang, J., Xie, J., and Xu, N. (2002). *Advanced Materials* 14 (13–14): 963–965.

57 Fu, C. and Guo, J. (2016). *Current Opinion in Chemical Engineering* 13: 53–62.

58 Eftekhari, A. and Kim, D.-W. (2017). *Journal of Materials Chemistry A* 5 (34): 17734–17776.

59 Deng, N., Kang, W., Liu, Y. et al. (2016). *Journal of Power Sources* 331: 132–155.

60 Jin, Z., Kai, X., and Hong, X. (2014). *Acta Chimica Sinica* 72 (1): 11.

61 Peled, E., Gorenshtein, A., Segal, M., and Sternberg, Y. (1989). *Journal of Power Sources* 26: 269–271.

62 Xiulei, J., Kyu, T., Lee, N., and Linda, F. (2009). *Nature Materials* 8 (6): 500–506.

63 Xin, S., Gu, L., Zhao, N.H. et al. (2012). *Journal of the American Chemical Society* 134 (45): 18510–18513.

64 Papandrea, B., Xu, X., Xu, Y. et al. (2016). *Nano Research* 9 (1): 240–248.

65 Lee, J.S., Kim, W., Jang, J., and Manthiram, A. (2017). *Advanced Energy Materials* 7 (5): 1601943.

66 Liang, J., Sun, Z.-H., Li, F., and Cheng, H.-M. (2016). *Energy Storage Materials* 2: 76–106.

67 Li, S., Bo, J., Zhai, X. et al. (2018). *ChemistrySelect* 3 (8): 2245–2260.

68 Tang, Q., Li, H., Zuo, M. et al. (2017). *Nano* 12 (2): 1750021.

69 Yu, M., Li, R., Wu, M., and Shi, G. (2015). *Energy Storage Materials* 1: 51–73.
70 Schipper, F., Vizintin, A., Ren, J. et al. (2015). *ChemSusChem* 8 (18): 3077–3083.
71 Li, X. (2014). *Frontiers in Energy Research* 2: 1.
72 Wu, D., Liu, J., Chen, J. et al. (2021). *Journal of Materials Chemistry A* 9: 5497–5506.
73 Ramakumar, S., Dasari, B., Ramesha, K., and Karkera, G. (2021). *Energy & Fuels* 35 (9): 8286–8294.
74 Hou, T.Z., Chen, X., Peng, H.J. et al. (2016). *Small* 12: 3283–3291.
75 Salem, H.A., Babu, G., Rao, C.V., and Arava, L. (2015). *Journal of the American Chemical Society* 137 (36): 11542.
76 Deng, C., Wang, Z., Feng, L. et al. (2020). *Journal of Materials Chemistry A* 8: 19704–19728.
77 Yu, M., Zhou, S., Wang, Z. et al. (2019). *Energy Storage Materials* 20: 98–107.
78 Song, Y., Cai, W., Kong, L. et al. (2020). *Advanced Energy Materials* 10 (11): 1901075.
79 Hong, X., Wang, R., Liu, Y. et al. (2020). *Journal of Energy Chemistry* 42: 144–168.
80 Yang, D.D., Hong, P.D., Hu, Y. et al. (2021). *Applied Surface Science* 552: 149424.
81 He, J., Bhargav, A., Asl, H.Y. et al. (2020). *Advanced Energy Materials* 10 (23): 2001017.
82 Wu, Y., Zhu, X., Li, P. et al. (2019). *Nano Energy* 59: 636–643.
83 Wu, H., Jiang, H., Yang, Y. et al. (2020). *Journal of Materials Chemistry A* 8 (29): 14498–14505.
84 Xia, G., Zheng, Z., Ye, J. et al. (2020). *Chemical Engineering Journal* 406: 126823.
85 Ye, Z., Jiang, Y., Li, L. et al. (2020). *Advanced Materials* 32 (32): 2002168.
86 Shi, Z., Sun, Z., Cai, J. et al. (2020). *Advanced Functional Materials* 31 (4): 2006798.
87 Wang, F., Li, J., Zhao, J. et al. (2020). *ACS Materials Letters* 2 (11): 1450–1463.
88 Zhou, G., Tian, H., Yang, J. et al. (2017). *Proceedings of the National Academy of Sciences of the United States of America* 114 (5): 840.
89 Lu, Y., Tu, Z., Shu, J., and Archer, L.A. (2015). *Journal of Power Sources* 279: 413–418.
90 Su, Y.-S. and Manthiram, A. (2012). *Nature Communications* 3 (1): 1–6.
91 Li, Z., Tang, L., Liu, X. et al. (2019). *Electrochimica Acta* 310: 1–12.
92 Song, Y., Zhao, S., Chen, Y. et al. (2019). *ACS Applied Materials & Interfaces* 11 (6): 5687–5694.
93 Han, P. and Manthiram, A. (2017). *Journal of Power Sources* 369: 87–94.
94 Gu, S., Sun, C., Xu, D. et al. (2018). *Electrochemical Energy Reviews* 1 (4): 599–624.
95 Gu, S., Qian, R., Jin, J. et al. (2016). *Physical Chemistry Chemical Physics* 18 (42): 29293–29299.
96 Xu, R.-c., Xia, X.-h., Wang, X.-l. et al. (2017). *Journal of Materials Chemistry A* 5 (6): 2829–2834.

97 Unemoto, A., Yasaku, S., Nogami, G. et al. (2014). *Applied Physics Letters* 105 (8): 083901.

98 Lau, K.C., Rago, N., and Chen, L. (2019). *Journal of the Electrochemical Society* 166 (12): A2570–A2573.

99 Fan, L.L., Deng, N.P., Yan, J. et al. (2019). *Chemical Engineering Journal* 369: 874–897.

100 Zhang, S., Ueno, K., Dokko, K., and Watanabe, M. (2015). *Advanced Energy Materials* 5 (16): 1500117.

101 Judez, X., Martinez-Ibanez, M., Santiago, A. et al. (2019). *Journal of Power Sources* 438: 226981–226985.

102 Yue, J., Yan, M., Yin, Y.X., and Guo, Y.G. (2018). *Advanced Functional Materials* 28 (38): 1707533.

103 Liu, Y., He, P., and Zhou, H. (2018). *Advanced Energy Materials* 8 (4): 1701602.

104 Camacho-Forero, L.E., Smith, T.W., Bertolini, S., and Balbuena, P.B. (2015). *Journal of Physical Chemistry C* 119 (48): 26828–26839.

105 Cheng, X.-B., Huang, J.-Q., and Zhang, Q. (2017). *Journal of the Electrochemical Society* 165 (1): A6058.

106 Cao, R., Wu, X., Lv, D. et al. (2015). *Advanced Energy Materials* 5 (16): 513–537.

107 Xiong, S., Xie, K., Diao, Y., and Hong, X. (2012). *Electrochimica Acta* 83: 78–86.

108 Zhang, L., Ling, M., Feng, J. et al. (2018). *Energy Storage Materials* 11: 24–29.

109 Cha, E., Patel, M.D., Park, J. et al. (2018). *Nature Nanotechnology* 13 (4): 337–344.

110 Liu, Y., Lin, D., Liang, Z. et al. (2016). *Nature Communications* 7 (1): 1–9.

111 Lin, Z., Liu, Z., Fu, W. et al. (2013). *Advanced Functional Materials* 23 (8): 1064–1069.

112 Akhtar, N., Sun, X., Akram, M.Y. et al. (2021). *Journal of Energy Chemistry* 52: 310–317.

113 Kzlaslan, A., Etinkaya, T., and Akbulut, H. (2020). *Advanced Materials Interfaces* 7 (20): 2001020.

114 Xiong, C., Ren, Y.X., Jiang, H.R. et al. (2019). *Journal of Energy Storage* 26: 101006.

115 Xu, R., Zhang, X.Q., Cheng, X.B. et al. (2018). *Advanced Functional Materials* 28 (8): 1705838.

116 Zhang, X., Wang, W., Wang, A. et al. (2014). *Journal of Materials Chemistry A* 2 (30): 11660–11665.

117 Eftekhari, A. (2017). *Sustainable Energy & Fuels* 1 (1): 14–29.

118 Liu, Y., Wang, J., Xu, Y. et al. (2014). *Journal of Materials Chemistry A* 2 (31): 12201–12207.

119 Koketsu, T., Paul, B., Wu, C. et al. (2016). *Journal of Applied Electrochemistry* 46 (6): 627–633.

120 Chen, Z., Zhao, Y., Mo, F. et al. (2020). *Small Structures* 1 (2): 2000005.

121 Abouimrane, A., Dambournet, D., Chapman, K.W., and Chupas, P.J. (2012). *Journal of the American Chemical Society* 134 (10): 4505–4508.

122 Cui, Y., Abouimrane, A., Sun, C.J. et al. (2014). *Chemical Communications* 50 (42): 5576–5579.

123 Zhou, X., Gao, P., Sun, S. et al. (2015). *Chemistry of Materials* 27 (19): 6730–6736.

124 Park, S., Park, J.S., and Kang, Y.C. (2018). *Journal of Materials Chemistry A* 6 (3): 1028–1036.

125 Li, X., Liang, J., Hou, Z. et al. (2015). *Advanced Functional Materials* 25 (32): 5229–5238.

126 Luo, C., Xu, Y., Zhu, Y. et al. (2013). *ACS Nano* 7 (9): 8003–8010.

127 Cui, Y., Abouimrane, A., Lu, J. et al. (2013). *Journal of the American Chemical Society* 135 (21): 8047–8056.

128 Zeng, L.C., Li, W.H., Jiang, Y., and Yu, Y. (2017). *Rare Metals* 36 (5): 1–26.

129 Jin, J., Tian, X., Srikanth, N. et al. (2017). *Journal of Materials Chemistry A* 5 (21): 10110–10126.

130 Ran, F. and Chen, S.W. (2019). *Advanced Nanomaterials for Electrochemical-Based Energy Conversion and Storage*, 69–114. Elsevier Inc.

131 Patil, A.M., Kumbhar, V.S., Chodankar, N.R. et al. (2016). *Journal of Colloid and Interface Science* 469: 257–262.

132 Liu, L., Hou, Y., Wu, X. et al. (2013). *Chemical Communications* 49 (98): 11515–11517.

133 Lee, S., Lee, H., Ha, N. et al. (2020). *Advanced Functional Materials* 30 (19): 2000028.

134 Xin, S., Yu, L., You, Y. et al. (2016). *Nano Letters* 16: 4560–4568.

135 Zeng, L., Zeng, W., Jiang, Y. et al. (2015). *Advanced Energy Materials* 5 (4): 1401377.

136 He, J., Chen, Y., Lv, W. et al. (2016). *ACS Energy Letters* 1 (1): 16–20.

137 Han, K., Liu, Z., Ye, H., and Dai, F. (2014). *Journal of Power Sources* 263: 85–89.

138 Kai, H., Zhao, L., Shen, J. et al. (2015). *Advanced Functional Materials* 25 (3): 455–463.

139 Lee, J.T., Kim, H., Oschatz, M. et al. (2015). *Advanced Energy Materials* 5 (1): 1–7.

140 Jiang, Y., Ma, X., Feng, J., and Xiong, S. (2015). *Journal of Materials Chemistry A* 3 (8): 4539–4546.

141 Choi, D.S., Yeom, M.S., Kim, Y.-T. et al. (2018). *Inorganic Chemistry* 57 (4): 2149–2156.

142 He, J., Lv, W., Chen, Y. et al. (2017). *Journal of Power Sources* 363: 103–109.

143 Tian, H., Tian, H., Wang, S. et al. (2020). *Nature Communications* 11 (1): 1–12.

144 Hu, J., Ren, Y., and Zhang, L. (2020). *Journal of Power Sources* 455: 227955.

145 Zhang, J., Li, Z., and Lou, X.W. (2017). *Angewandte Chemie International Edition* 56 (45): 14107–14112.

146 Xu, H., Chen, G., Jin, R. et al. (2014). *RSC Advances* 4 (17): 8922–8929.

147 Wu, F., Lee, J.T., Xiao, Y., and Yushin, G. (2016). *Nano Energy* 27: 238–246.

148 Bachman, J.C., Muy, S., Grimaud, A. et al. (2016). *Chemical Reviews* 116 (1): 140–162.

149 Cheng, Z., Pan, H., Zhong, H. et al. (2018). *Advanced Functional Materials* 28 (38): 1707597.

150 Li, X., Liang, J., Luo, J. et al. (2019). *Advanced Materials* 31 (17): 1808100.

151 Zhang, Q., Liu, K., Ding, F., and Liu, X. (2017). *Nano Research* 10 (12): 4139–4174.

152 Xu, L., Tang, S., Cheng, Y. et al. (2018). *Joule* 2 (10): 1991–2015.

153 Lee, J.T., Kim, H., Nitta, N. et al. (2014). *Journal of Materials Chemistry A* 2 (44): 18898–18905.

154 Wang, W.P., Zhang, J., Yin, Y.X. et al. (2020). *Advanced Materials* 32 (23): 2000302.

155 Zhang, Y., Manaig, D., Freschi, D.J., and Liu, J. (2021). *Energy Storage Materials* 40: 166–188.

156 Zhang, X., Jiao, S., Tu, J. et al. (2019). *Energy & Environmental Science* 12 (6): 1918–1927.

157 Zhang, J., Yin, Y.X., You, Y. et al. (2014). *Energy Technology* 2: 757–762.

158 Zhang, Y., Lu, W., Zhao, P. et al. (2021). *Electrochimica Acta* 388: 138621.

159 Zhang, Y., Lu, W., Zhao, P. et al. (2021). *Carbon* 173: 11–21.

160 He, J., Chen, Y., Lv, W. et al. (2016). *ACS Nano* 10 (9): 8837–8842.

161 Xu, J., Xin, S., Liu, J.W. et al. (2016). *Advanced Functional Materials* 26 (21): 3580–3588.

162 Sun, F., Zhang, B., Tang, H. et al. (2018). *Journal of Materials Chemistry A* 6 (21): 10104–10110.

163 Li, S., Han, Z., Hu, W. et al. (2019). *Nano Energy* 60: 153–161.

164 Ding, N., Chen, S.F., Geng, D.S. et al. (2015). *Advanced Energy Materials* 5 (8): 1401999.

165 He, J., Lv, W., Chen, Y. et al. (2017). *ACS Nano* 11 (8): 8144–8152.

166 Yin, H., Yu, X.-X., Yu, Y.-W. et al. (2018). *Electrochimica Acta* 282: 870–876.

167 Huang, D., Li, S., Xiao, X. et al. (2017). *Journal of Power Sources* 371: 48–54.

168 Li, Y., Zhang, Y., Xu, Q. et al. (2019). *ChemSusChem* 12 (6): 1196–1202.

169 Lu, G., Ye, C., Li, W. et al. (2021). *Frontiers in Chemistry* 9: 377.

170 Zhang, S.S. (2013). *Frontiers in Energy Research* 1: 10.

171 Yang, H., Chen, J., Yang, J., and Wang, J. (2020). *Angewandte Chemie* 132 (19): 7374–7386.

172 Chen, S. and Zhang, J. (2020). *Dalton Transactions* 49 (29): 9929–9934.

173 Xing, M., Zhao, Z., Zhang, Y. et al. (2020). *Materials Today Energy* 12: 100534.

174 Xie, C., Liu, Y., Lu, W. et al. (2019). *Energy & Environmental Science* 12 (6): 1834–1839.

175 Ma, J., Liu, M., He, Y., and Zhang, J. (2021). *Angewandte Chemie* 133 (23): 12744–12755.

176 Skarstad, P.M. and Schmidt, C.L. (1993). *Journal of Power Sources* 43: 111–118.

177 Wang, Y., Sun, Q., Zhao, Q. et al. (2011). *Energy & Environmental Science* 4 (10): 3947–3950.

178 Li, K., Lin, B., Li, Q. et al. (2017). *ACS Applied Materials & Interfaces* 9 (24): 20508–20518.

179 Zhao, Q., Lu, Y., Zhu, Z. et al. (2015). *Nano Letters* 15 (9): 5982–5987.

180 Zhang, Q., Wu, Z., Liu, F. et al. (2017). *Journal of Materials Chemistry A* 5 (29): 15235–15242.

181 Su, Z., Ling, H.Y., Li, M. et al. (2020). *Carbon Energy* 2 (2): 265–275.
182 Zhang, Q., Zeng, Y.-H., Ye, S.-H., and Liu, S. (2020). *Journal of Power Sources* 463: 228212.
183 Su, Z., Wei, Z., Lai, C. et al. (2018). *Energy Storage Materials* 14: 129–135.
184 Li, K., Chen, S., Chen, S. et al. (2019). *Nano Research* 12 (3): 549–555.
185 Sun, C., Shi, X., Zhang, Y. et al. (2020). *ACS Nano* 14 (1): 1176–1184.
186 Wang, F., Liu, Z., Yang, C. et al. (2020). *Advanced Materials* 32 (4): 1905361.
187 Fang, C., Wang, X., and Meng, Y.S. (2019). *Trends in Chemistry* 1 (2): 152–158.
188 Zhang, X.Q., Cheng, X.B., and Zhang, Q. (2018). *Advanced Materials Interfaces* 5 (2): 1701097.
189 Li, K., Hu, Z., Ma, J. et al. (2019). *Advanced Materials* 31 (33): 1902399.
190 Lai, Q., Zhang, H., Li, X. et al. (2013). *Journal of Power Sources* 235: 1–4.
191 Chang, Z., Wang, X., Yang, Y. et al. (2014). *Journal of Materials Chemistry A* 2 (45): 19444–19450.
192 Xi, X., Li, X., Wang, C. et al. (2017). *Journal of Energy Chemistry* 26 (4): 639–646.
193 Etacheri, V., Marom, R., Elazari, R. et al. (2011). *Energy & Environmental Science* 4 (9): 3243–3262.
194 Li, H. (2019). *Joule* 3 (4): 911–914.
195 Li, B., Wang, Y., Jiang, N. et al. (2020). *Nano Energy* 72: 104727.
196 Wu, W., Wang, M., Wang, J. et al. (2020). *Nano Energy* 74: 104867.
197 Hewavitharana, I.K., Ding, Y., Ng, K.S., and Deng, D. (2021). *Chemical Engineering Science* 236: 116480.
198 Jung, S.-K., Kim, H., Cho, M.G. et al. (2017). *Nature Energy* 2 (2): 1–9.
199 Paul, P.P., McShane, E.J., Colclasure, A.M. et al. (2021). *Advanced Energy Materials* 11 (17): 2100372.
200 Lin, D., Liu, Y., and Cui, Y. (2017). *Nature Nanotechnology* 12 (3): 194.
201 Jie, Y., Ren, X., Cao, R. et al. (2020). *Advanced Functional Materials* 30 (25): 1910777.
202 Qian, J., Henderson, W.A., Xu, W. et al. (2015). *Nature Communications* 6 (1): 1–9.
203 Kamaya, N., Homma, K., Yamakawa, Y. et al. (2011). *Nature Materials* 10 (9): 682–686.
204 Li, N.W., Yin, Y.X., Yang, C.P., and Guo, Y.G. (2016). *Advanced Materials* 28 (9): 1853–1858.
205 Lin, D., Liu, Y., Liang, Z. et al. (2016). *Nature Nanotechnology* 11 (7): 626–632.
206 Cheng, X.-B., Zhao, C.-Z., Yao, Y.-X. et al. (2019). *Chem* 5 (1): 74–96.
207 Zhao, W., Ji, Y., Zhang, Z. et al. (2017). *Current Opinion in Electrochemistry* 6 (1): 84–91.
208 Patten, H.E. and Mott, W.R. (2002). *The Journal of Physical Chemistry* 8 (3): 153–195.
209 Meibuhr, S.G. (1971). *Journal of the Electrochemical Society* 118 (8): 1320.
210 Whittingham, M.S. and Gamble, F.R. Jr., (1975). *Materials Research Bulletin* 10 (5): 363–371.

211 Gachot, G., Grugeon, S., Armand, M. et al. (2008). *Journal of Power Sources* 178 (1): 409–421.

212 Miao, R., Yang, J., Xu, Z. et al. (2016). *Scientific Reports* 6 (1): 1–9.

213 Zhang, H., Eshetu, G.G., Judez, X. et al. (2018). *Angewandte Chemie International Edition* 57 (46): 15002–15027.

214 Li, F., He, J., Liu, J. et al. (2020). *Angewandte Chemie International Edition* 60 (12): 6600–6608.

215 Qi, S., He, J., Liu, J. et al. (2021). *Advanced Functional Materials* 31 (11): 2009013.

216 Amine, R., Liu, J., Acznik, I. et al. (2020). *Advanced Energy Materials* 10 (25): 2000901.

217 Nilsson, V., Kotronia, A., Lacey, M. et al. (2019). *ACS Applied Energy Materials* 3 (1): 200–207.

218 Ren, X., Chen, S., Lee, H. et al. (2018). *Chem* 4 (8): 1877–1892.

219 Cao, X., Xu, Y., Zhang, L. et al. (2019). *ACS Energy Letters* 4 (10): 2529–2534.

220 Jiao, S., Ren, X., Cao, R. et al. (2018). *Nature Energy* 3 (9): 739–746.

221 Li, Q., Jiao, S., Luo, L. et al. (2017). *ACS Applied Materials & Interfaces* 9 (22): 18826–18835.

2
Electrode–Electrolyte Interphase

2.1 Introduction

As the "Holy Grail" of lithium batteries, lithium metal batteries have received great attention due to their high energy density and high working voltage. However, the continuous growth of lithium dendrites and the side reaction of electrodes with electrolytes due to high reactivity of lithium greatly hinder the development of lithium metal batteries [1–3]. Many efforts have been devoted to addressing these problems; strategies such as constructing three-dimensional electrodes, employing modified electrode materials, and finely optimizing the electrolyte composition have already been extensively studied [4–7]. Among them, the regulation of electrode–electrolyte interphase by optimizing the electrolytes has been considered the effective method with low cost. Many studies have demonstrated that the electrode–electrolyte interphase plays an important role in the reversible operation of Li–ion batteries [8–10].

In this section, we give an overview of the concept and roles of electrode–electrolyte interphase on both anode and cathode in battery performance. The composition and nature of solid electrolyte interphase (SEI) and cathode electrolyte interphase (CEI) for stable lithium metal batteries are introduced. The types and modification strategies of SEI/CEI regulation are also summarized in detail. The guidelines for the rational design of electrode–electrolyte interphase are proposed.

2.2 Solid Electrolyte Interphase

2.2.1 Concept and Roles

Due to high reactivity of Li with most nonaqueous electrolytes, during the cycles, a protection layer is generated between the electrode and electrolyte. Such a protection layer is usually electrically insulating but allows Li^+ diffusion. The layer halts decomposition of the electrolyte while the battery is running continuously so that the commercialization of lithium–ion battery can be realized [11]. For lithium metal batteries, one of the most important problems that hinder their application is safety concerns caused by the uniform growth of Li dendrites [12]. The published works have demonstrated that the Li dendrite growth is closely related to the nature of

Liquid Electrolyte Chemistry for Lithium Metal Batteries: Design, Mechanisms, Strategies,
First Edition. Jianmin Ma.

the SEI. The SEI with uneven structure will lead to uneven distribution of surface current on the anode and thus lead to uneven deposition of lithium, which will produce lithium dendrites [13]. Thus, the regulation of SEI is of great importance for the inhibition of Li dendrites. Besides, as the protection layer on the anode, the SEI with electrical insulation could also protect the electrolyte from the continuous corrosion of the electrode; thus, the strategies of SEI with unique structure are very important [14, 15]. Taking into account these considerations, in recent years, many efforts have been devoted to the rational design of electrolytes that could generate the SEI with the robust and uniform structure that could inhibit the Li dendrites and protect the electrolyte simultaneously.

2.2.2 Types and Modification Strategies

SEI is generated between the electrode and electrolyte because of thermodynamic instability of lithium in most nonaqueous solvents used in lithium batteries, so the components of electrolyte exert high influence on the nature and properties of SEI [16]. The compatibility of electrolytes with the selected electrodes is the first principle in the design of a new battery system. Published works have demonstrated that an ideal electrolyte should have a high dielectric constant, which is conducive to the dissolution of lithium salts, and low viscosity, which is favorable for ion migration. For this consideration, the commonly used electrolytes are usually composed of a mixture of two or more carbonates: cyclic carbonates, such as ethylene carbonate (EC), propylene carbonate (PC), and linear carbonates, such as dimethyl carbonate (DMC), ethyl methyl carbonate (EMC), and so on. Cyclic carbonates with high dielectric constants could facilitate dissociation of lithium salts, whereas linear carbonates with low viscosity could reduce the viscosity of electrolytes [17, 18]. In lithium metal batteries, growth of lithium dendrites is a key problem that cannot be ignored; therefore, a reasonable design of electrolytes with optimized components that are not only compatible with electrodes but also can inhibit the growth of Li dendrites is of great importance. For a better understanding of the relationship of SEI components with the solvated Li kinetics, Huang and his coworkers reported a decrease in the Li interfacial transfer barrier by SEI rich in inorganics species. The electrochemical impedance spectroscopy (EIS) was first conducted to probe the formation process of SEI on the Li metal. Then, the chemical composition of SEI formed in three different electrolyte systems was characterized by X-ray photoelectron spectroscopy (XPS) (Figure 2.1). It was found that the easy cleavage of C–F in bis(fluorosulfonyl)imide (FSI) than bis(trifluoromethane)sulfonimide (TFSI) leads to the formation of SEI with abundant LiF, while high anion contents in Li solvation structure in high-concentration electrolytes lead to the formation of inorganic species in SEI. Besides, the EIS and Tafel plot tests proved that the SEI with more inorganic species led to fast charge–transfer behavior during the solvated Li interfacial diffusion and therefore resulted in uniform deposition of Li [19].

Besides the components of electrolyte, anions could also greatly influence the properties of SEI. From the viewpoint of thermodynamics, Li salts will be solvated by solvents to form the solvation structure, in which case the Li^+ and anions will be separated to some extent, the solvents and anions will consist of the main part

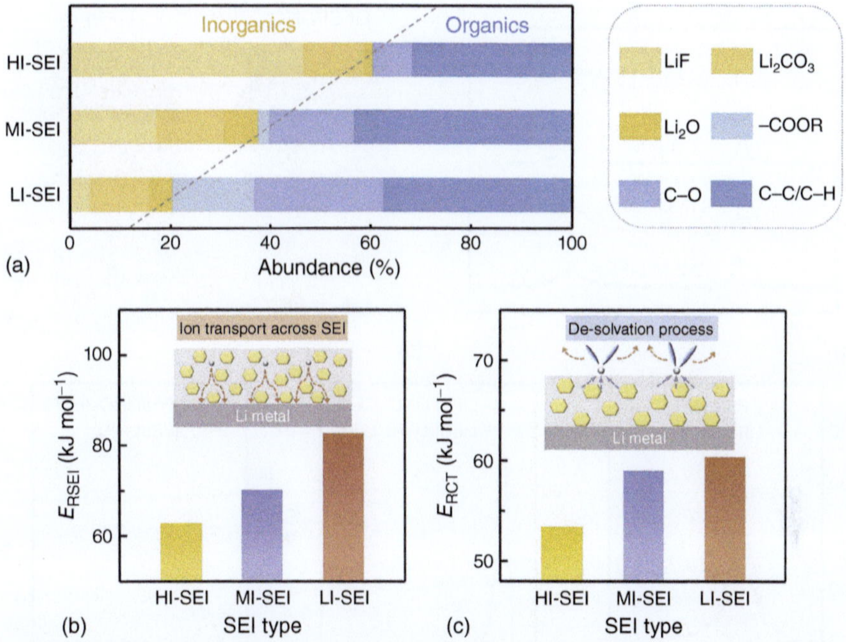

Figure 2.1 (a) Abundance of various organic/inorganic components in HI-SEI, MI-SEI, and LI-SEI calculated according to the corresponding fitted peak areas and the atomic ratios determined by XPS. Comparisons of activation energy for decoupled processes of (b) ion transport across SEI (ERSEI) and (c) de-solvation (ERCT) with regard to HI-SEI, MI-SEI, and LI-SEI by fitting the RSEI and RCT obtained at various temperatures (−20 to 30 °C) with the Arrhenius equation. Source: Reproduced with permission Xu et al. [19].

of the Li solvation structure and will greatly influence the migration of Li^+ in the solvent and electrode. Specifically, when Li^+ diffuses to the anode to be reduced, the solvents and anions in the Li solvation structure will also be reduced to form the SEI. In this case, the structure and properties of SEI are also largely affected by the solvation of lithium. Zhang and his coworkers studied the interrelation among various anions in the Li solvation structure by introducing $LiNO_3$ into the model system in dimethoxyethane (DME). Upon the addition of $LiNO_3$, the oxidation decomposition potentials of LiFSI electrolyte increased from 3.3 to 4.3 V. The much-enlarged oxidation window of LiFSI electrolyte was attributed to the increased ion–dipole interaction between Li ions and the DME solvent. By combining MD simulations with the ^{17}O NMR spectra (Figure 2.2), they also demonstrated that the Li solvation structure was greatly changed by the NO_3 anions; the upfield shift of O_{DME} peaks and O_{FSI} in the LiFSI electrolyte after addition of NO_3^- demonstrated that NO_3^- was successfully introduced to the solvation sheath, while the FSI^- anions decomposed completely. By in-depth XPS, the new species of SEI in the lower state of oxidation were found and complete decomposition of FSI^- was demonstrated. In summary, they demonstrated the feasibility of modifying the composition of SEI by introducing anions into solvents to regulate the solvation structure of Li^+ without significantly reducing ionic conductivity [20].

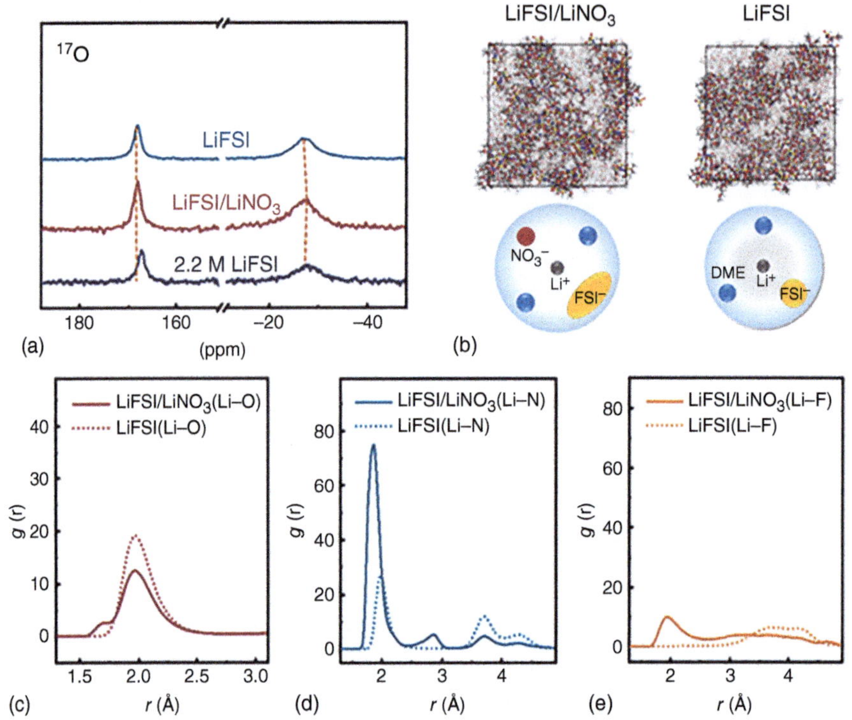

Figure 2.2 (a) Natural abundance ^{17}O NMR spectra of various electrolytes measured at 50 °C. (b) Top panel: snapshots of the MD simulation boxes of LiFSI/$LiNO_3$ and LiFSI electrolytes. MD simulations of (c) Li–O, (d) Li–N, and (e) Li–F radial distribution functions, *g*(r). Source: Reproduced with permission Zhang et al. [20].

Under similar conditions, Wang and his coworkers reported the regulation of sulfone electrolytes with $LiNO_3$ for lithium batteries to suppress the growth of Li dendrites. Different from the abovementioned studies, in order to match the high-voltage cathode, they introduced $LiNO_3$ into the cyclic sulfone solvent with higher oxidative stability. The FTIR spectra and molecular dynamics (MD) simulations (Figure 2.3) demonstrated that the interactions of $TFSI^-$ with Li^+ were enhanced by the introduced NO_3. The corresponding shift of $TFSI^-$ reduction potential to high voltage further demonstrated the stronger interrelation of Li^+ with TFSI. Besides, in-depth XPS with Ar^+ sputtering for different depths demonstrated that NO_3 promoted the cleavage of the C—F bond in TFSI and therefore resulted in the formation of SEI with more LiF and LiN_xO_y. The LiF with high interface energy and low electron conductivity could lead to the parallel direction diffusion of Li to avoid the growth of Li dendrites and prevent electron leakage of SEI [21].

Introducing self-sacrificial additives is another effective and economical method for regulating the Li solvation structure to generate the SEI with desired properties. From the previous studies, we can conclude that additives with higher reduction potentials will be reduced on the anode earlier than solvents during the Li plating process to form an ideal SEI with designed properties. By a rational design of

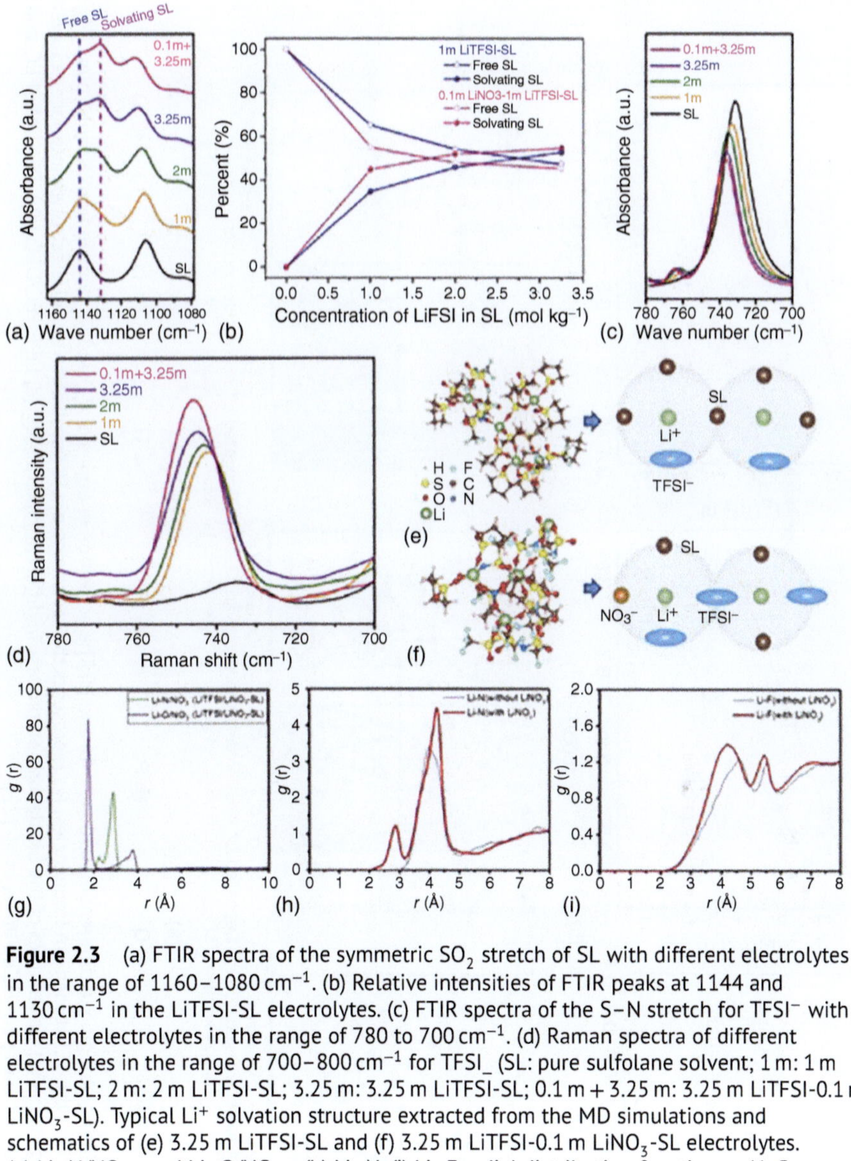

Figure 2.3 (a) FTIR spectra of the symmetric SO_2 stretch of SL with different electrolytes in the range of 1160–1080 cm^{-1}. (b) Relative intensities of FTIR peaks at 1144 and 1130 cm^{-1} in the LiTFSI-SL electrolytes. (c) FTIR spectra of the S–N stretch for $TFSI^-$ with different electrolytes in the range of 780 to 700 cm^{-1}. (d) Raman spectra of different electrolytes in the range of 700–800 cm^{-1} for TFSI_ (SL: pure sulfolane solvent; 1 m: 1 m LiTFSI-SL; 2 m: 2 m LiTFSI-SL; 3.25 m: 3.25 m LiTFSI-SL; 0.1 m + 3.25 m: 3.25 m LiTFSI-0.1 m $LiNO_3$-SL). Typical Li^+ solvation structure extracted from the MD simulations and schematics of (e) 3.25 m LiTFSI-SL and (f) 3.25 m LiTFSI-0.1 m $LiNO_3$-SL electrolytes. (g) Li–N/NO_3^- and Li–O/NO_3^-, (h) Li–N, (i) Li–F radial distribution functions, *g*(r). Source: Reproduced with permission Fu et al. [21].

electrolyte systems with suitable additives, regulation of the SEI with desired species could be effectively achieved. In a report by Li and his coworkers, lithium trifluoroacetate (LiTFA) was induced into 1,2-dimethoxyethane (DME)/FEC electrolyte to significantly regulate the Li^+ solvation shell and prevent the uncontrollable growth of Li dendrites. The lower lowest unoccupied molecular orbital (LUMO) energy of the as-used LiTFA additive than that of DME and FEC suggested that LiTFA will be reduced earlier on Li metal, thus can produce stable and uniform SEI with

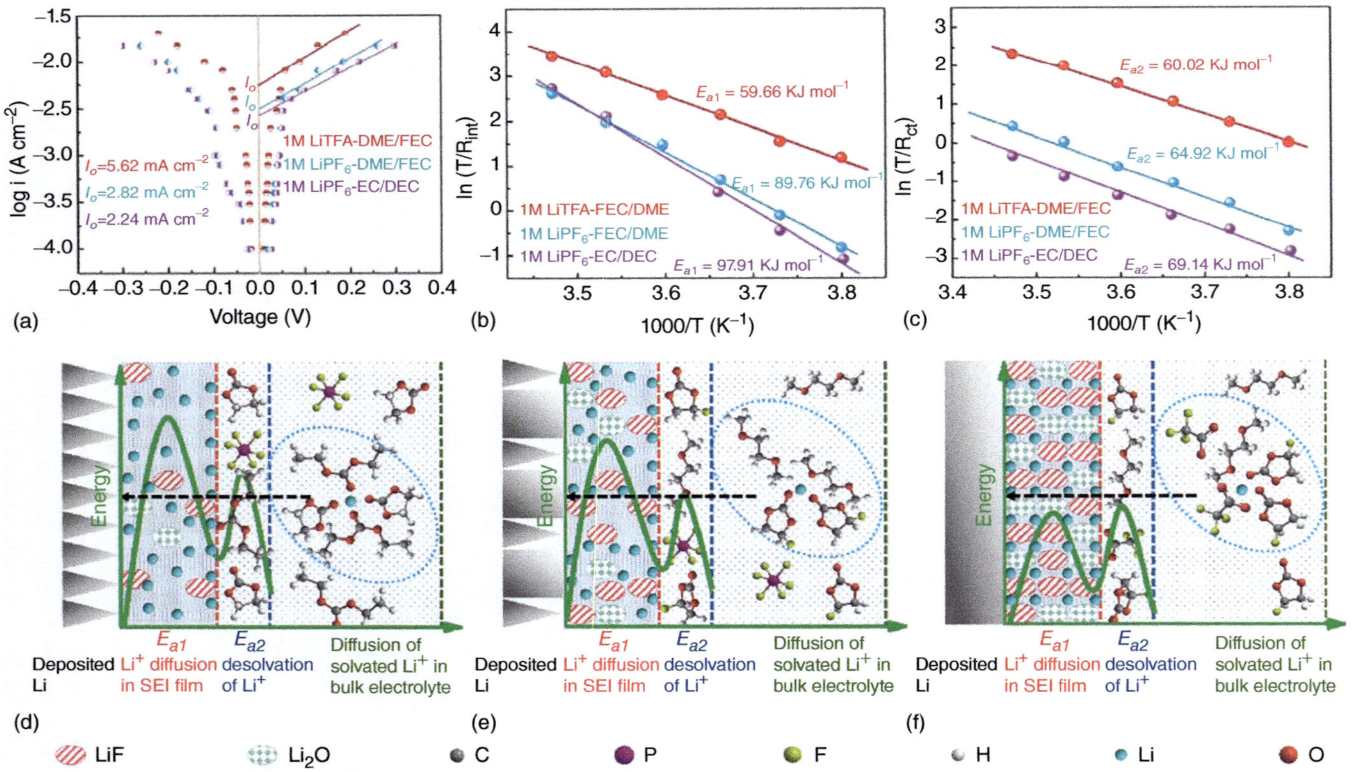

Figure 2.4 (a) Tafel plots for Li plating/stripping in different electrolytes. (b) and (c) The activation energies of R_{int} and R_{ct} derived from Nyquist plots. R_{int} represents the resistance of Li^+ across the SEI at medium frequencies; R_{ct} represents the resistance of Li^+ desolvation at SEI/electrolyte interface at lower frequencies (also denoted as charge transfer). Schematics of the Li^+ deposition process in different electrolytes. (d) $LiPF_6$-EC/DEC, (e) $LiPF_6$-DME/FEC, and (f) LiTFA-DME/FEC. Source: Reproduced with permission Wang et al. [22].

abundant LiF and LiO_2, which can be demonstrated by the results of XPS. Such SEI with abundant inorganic species can effectively reduce the Li^+ diffusion energy barrier, boost the ion transfer kinetics, and lower the nucleation overpotential of Li^+ (Figure 2.4). It was found that the Li^+ solvation sheath was regulated by TFA^- because of the strong coordination of carbonyl groups (C=O) in TFA^- with Li^+, and thereby the desolvation kinetics of Li was improved [22].

Based on the advantages of SEI with abundant inorganic species, Li and his coworkers reported on the use of liquid polydimethylsiloxane (PDMS) terminated with $-OCH_3$ groups (PDMS-OCH_3) as a graftable additive for a carbonate-based electrolyte to suppress anode dendrites. The as-used additive could facilitate the formation of a new type of SEI with a conformal hybrid structure, in which the LiF and Li-Si-O-based moieties were increased and the grafting of PDMS macromolecules on the surface of oxidic Li was enhanced. The decreased carbon signal implied that the decomposition of carbonate electrolyte was mitigated effectively, while the increasing F content and the fluorinated PDMS fragments suggested that SEI featured a hybrid structure with organic and inorganic components. The EIS displayed that the addition of PDMS did not decrease the diffusion of Li^+. The voltage curves of Li/Cu asymmetric cells demonstrated that the addition of PDMS-OCH_3 can effectively regulate the Li plating and stripping kinetics (Figure 2.5) [23]. Sun and his coworkers introduced octaphenyl polyoxyethylene as an electrolyte additive for the construction of a complex layer on lithium surface to suppress the growth of Li dendrites and promote the formation of robust SEI with cross-linked polymers. Because of the complexation of Li^+ with lithiophilic EO functional groups of the additives, Li deposition was well regulated; besides, the additives enhanced the stability of the electrolyte and acted as a template to promote the formation of SEI with stable structure; thus improved Li^+ transport and reduced the consumption of the electrolyte (Figure 2.6) [24].

Wang and his coworkers employed high-modulus LiF–organic bilayer interphase on high-capacity microsized Si, Al, and Bi anodes for high-performance batteries. To address the problems of fast degradation and low CE of alloy anodes, the SEI with high interfacial energy and low adhesion was designed to accommodate the volume change of alloy anodes during cycles. The morphology of Si anodes and the MD simulations demonstrated that thinner SEI with abundant high-modulus LiF uniformly could effectively suppress the pulverization of Li_xSi (Figure 2.7) [25]. Besides the SEI with high LiF, the nitriding interfaces for the regulation of Li deposition were studied by Guo and coworkers. Herein, $LiNO_3$ was used to modify the organic phosphate electrolytes to construct a compatible interface between OPEs and the Li metal anode. The deposition process of Li on the OPEs/MOPEs was studied for the first time. The source of the incompatibility between OPEs and Li was found to be instability of OPEs to Li and the formation of inhomogeneous SEI. By the introduction of $LiNO_3$-modified MOPEs, the nitriding interface was successfully constructed on the interface. The XPS proved that after the introduction of additives, the LiN_xO_y peaks appeared and the intensity of Li_3N increased; the Li_3N displayed high Li^+ conductivity and effectively decreased the interfacial

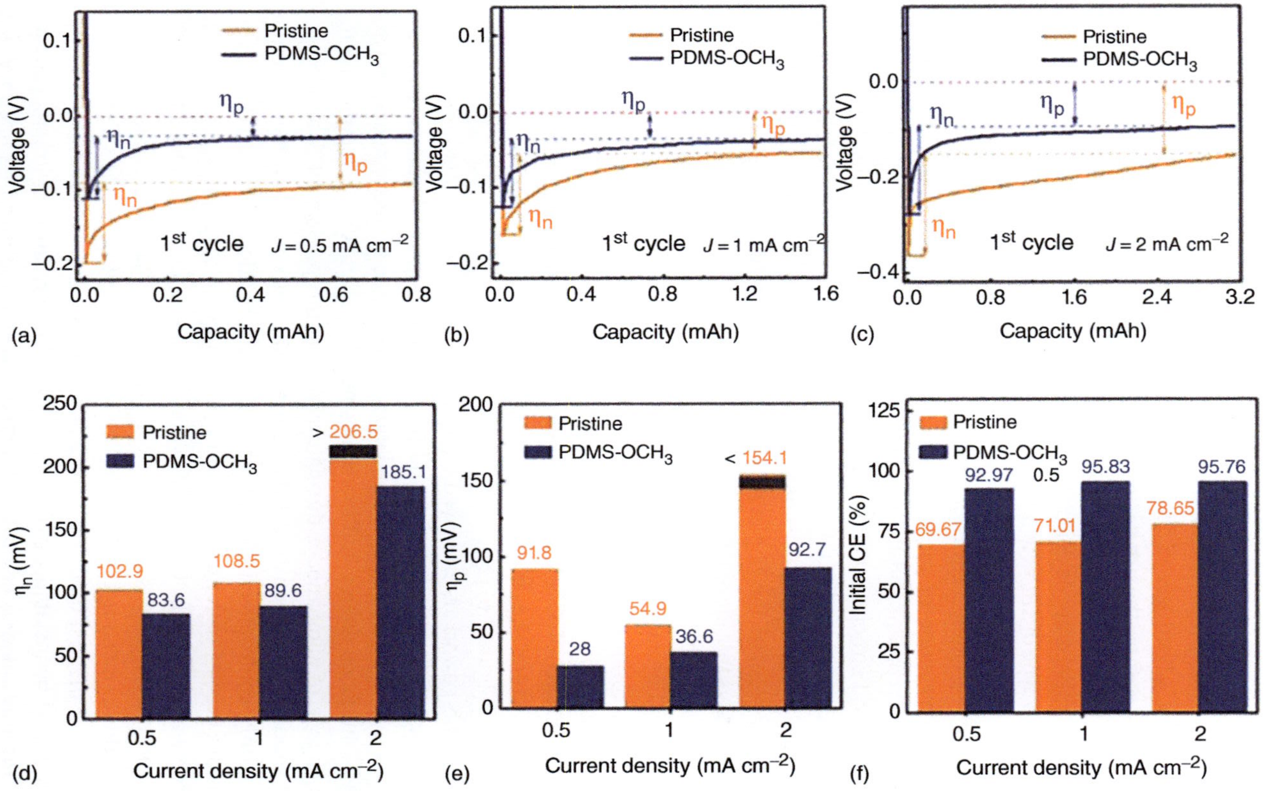

Figure 2.5 Comparison of voltage curves of Li/Cu asymmetric cells with and without PDMS addition during initial Li plating on Cu foil at (a) 0.5, (b) 1, and (c) 2 mA cm^{-2}. Nucleation and plateau overpotentials (η_n and η_p) are denoted. Comparison of (d) η_n values, (e) η_p values, and (f) initial CE values for pristine and modified electrolyte systems at 0.5, 1, and 2 mA cm^{-2}. Source: Reproduced with permission Meng et al. [23].

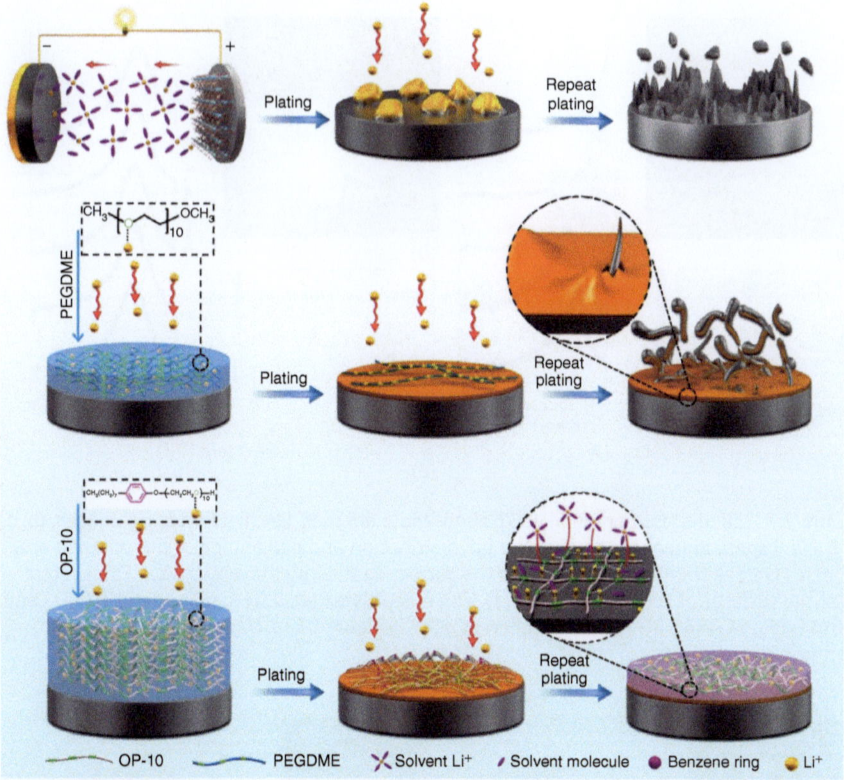

Figure 2.6 Schematic illustration of lithium deposition with and without various additives. Source: Dai et al. [24].

resistance (Figure 2.8). The as-generated Li_3N-rich SEI contributed to the uniform deposition of Li and better Li^+ transportation [26].

Ma and coworkers employed potassium perfluorinated sulfonates as multifunctional additives for lithium metal batteries with a high-voltage cathode. This work proposed the multifactor principle for the electrolyte additive molecular design (EAMD); unique insights into the modification mechanism of ionic electrolyte additives for battery on both anode and cathode were offered. Based on the reported studies, the advantages of LiF-enriched SEI and selected cations shielding mechanisms were achieved with only one additive. Related characterizations demonstrated that K^+ at low concentrations has a lower reduction potential than Li^+ to be absorbed on the surface of dendrites to suppress the growth of Li dendrites. Besides, the additive-regulated Li solvation layer enriched the LiF-rich SEI and effectively lowered the Li desolvation energy barrier (Figure 2.9). Based on the multifactor principle for the EAMD, a series of works have been reported by Ma and coworkers, and the principle of the EAMD has been demonstrated to be a useful guideline for further design of electrolytes to regulate the SEI for Li metal batteries (Figure 2.9) [27, 28].

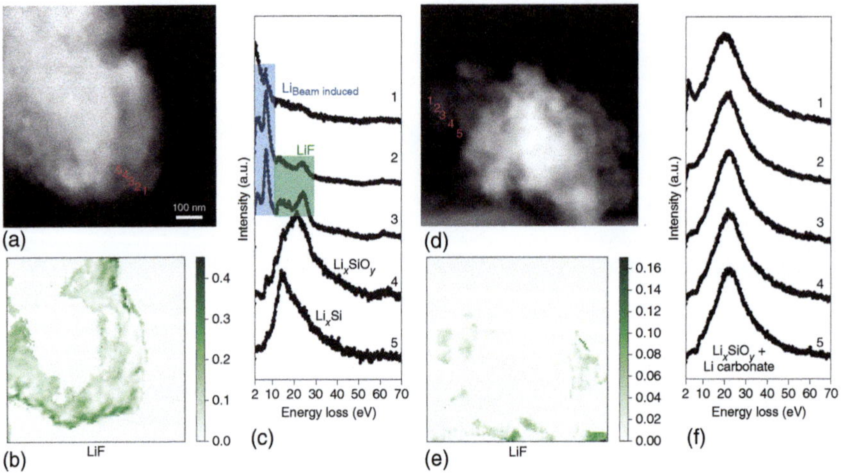

Figure 2.7 LiF distribution on Si. (a, d) High-angle annular dark-field imaging images. (b, c, e, f) EELS spectral images showing the LiF distribution and percentage of the scanned area (b, e), as well as five (labeled 1–5 from the surface to the inner layers) typical EEL spectra near the surfaces of the Si particles (c, f). The electrolytes are 2.0 M $LiPF_6$-mixTHF (a–c) and 1.0 M $LiPF_6$-EC-DMC (d–f). Source: Chen et al. [25], figure 05 (p. 07)/with permission of Springer Nature.

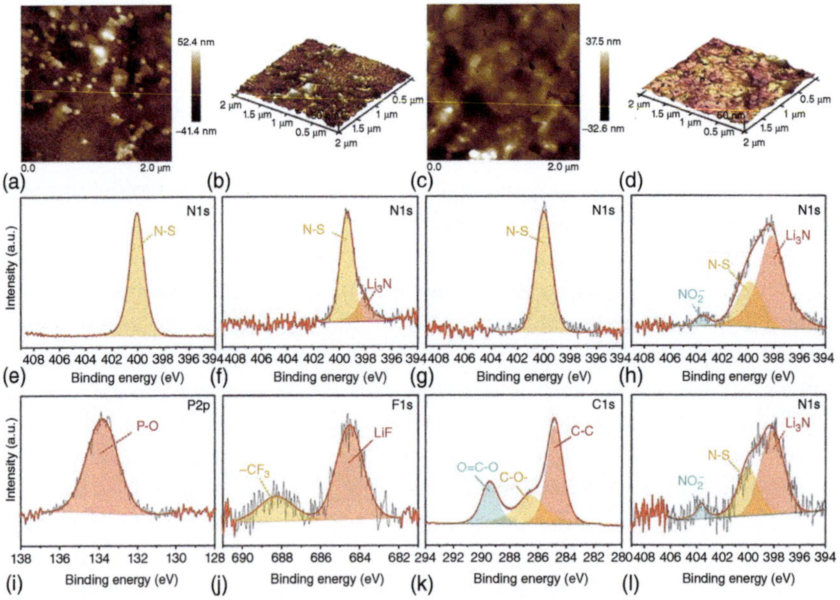

Figure 2.8 Morphology and composition of Li surfaces. 2D and 3D AFM images of Li surfaces after formation in (a, b) OPEs and (c, d) MOPEs. N 1s XPS spectra of lithium anodes after being immersed in (e) OPEs and (f) MOPEs for 4 h. (g) N 1s XPS spectra for OPEs and (h) N 1s, (i) P 2p, (j) F 1s, (k) C 1s spectra for MOPEs after formation process. (l) N 1s spectra for MOPEs after 150 cycles. Source: Tan et al. [26]/with permission of John Wiley and Sons.

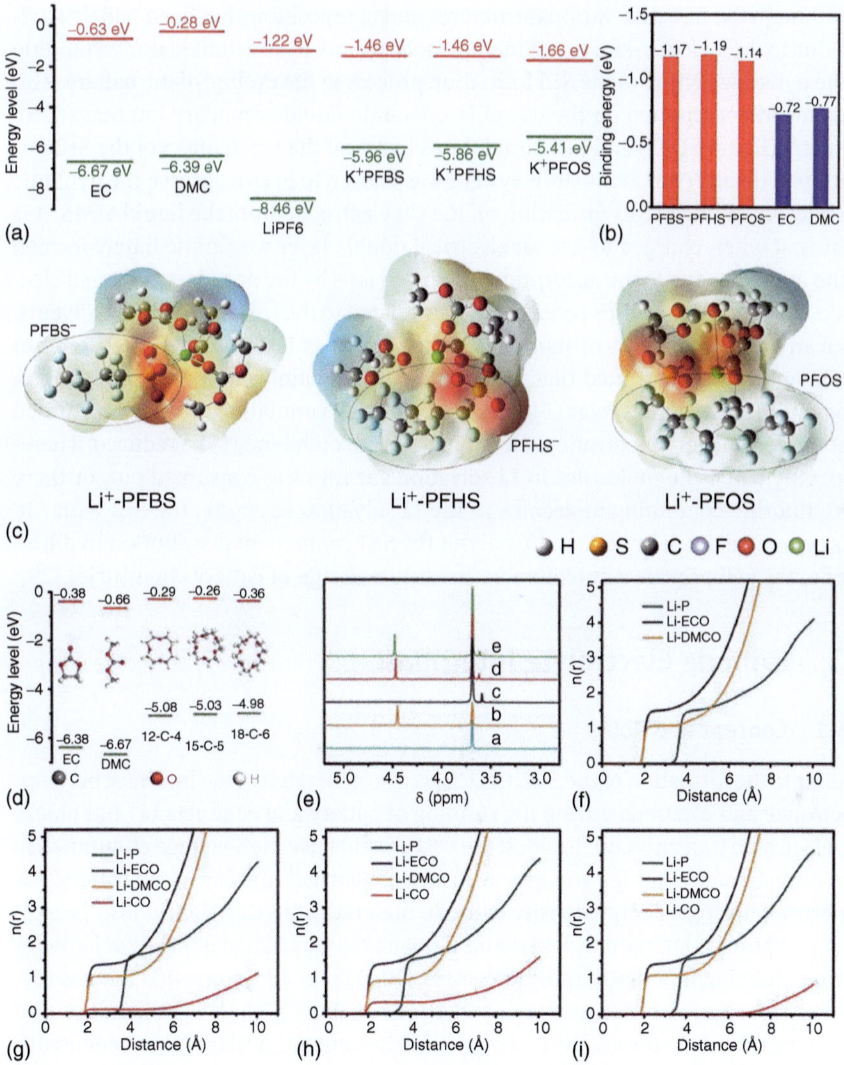

Figure 2.9 Theoretical calculation of the electrolyte with potassium perfluorinated sulfonates additives and the blank electrolyte. (a) The LUMO and HOMO energy values of the solvents, lithium salt, and additives, obtained by DFT calculation. Source: Reproduced with permission Qi et al. [27]. (b) The calculated binding energy with Li^+ ion of additives and solvent molecules. Snapshot of Li ion's first solvation shell with K^+PFBS, K^+PFHS, and K^+PFOS. Source: Reproduced with permission Wang et al. [28]. (c) Theoretical calculation of highest occupied–lowest unoccupied Mo (HOMO–LUMO) energy of EC, DMC, and additives. (d) Theoretical calculation of highest occupied–lowest unoccupied MO (HOMO–LUMO) energy of EC, DMC, 12-C-4, 15-C-5, and 18-C-6. (e) 1H NMR spectroscopy of eletrolyte before and after three cycle tests (a: DMC, b: EC+DMC, c: 15-C-5+DMC, d: 2.0 wt% 15-C-5 electrolyte before test, e: 2.0 wt% 15-C-5 electrolyte after test). The cumulative coordination number $n(r)$ of Li^+ in (f) blank, (g) with 12-C-4 additive, (h) with 15-C-5 additive, and (i) with 18-C-6 additive electrolyte system.

Although the SEI with various structures and compositions has been widely studied, due to the lack of in situ experimental tools, we still have a limited understanding of the dynamic picture of the SEI formation process in the cycling of the battery. Zhu and coworkers reported on the use of in operando liquid secondary ion mass spectrometry (liquid SIMS) and MD simulations to reveal the mechanism of the SEI formation in detail. The LiFSI–DME system was chosen to in situ observe the dynamic chemical mapping of SEI formation on the Cu electrode. From the liquid SIMS, it is clear that when charged to 1 V, an electrical double layer was immediately formed at the interface, due to the adsorption of lithium ions by the negatively charged electrode. The increase in DME concentration is related to the solvation of Li by solvents. Such an interface consists of abundant Li^+ and solvated DME, but a limited amount of FSI anions demonstrated that the inner SEI was mainly composed of reduction products of solvent molecules (Figure 2.10). The MD simulations also demonstrated that the solvents in the solvation sheath had a higher tendency to be reduced. Therefore, only when the molecules in Li solvation sheath were consumed out, or there were fluorine-containing molecules in the Li solvation structure, the SEI with LiF was generated. Such understanding about the SEI component distribution in different depths will provide a guideline for the future design of battery chemistries [29].

2.3 Cathode Electrolyte Interphase

2.3.1 Concept and Roles

Similar to the formation of the SEI, the CEI is also generated in the interface between electrolyte and electrode during the running of battery and conducts Li^+ but blocks the electronic transmission. Besides, the CEI should have high-voltage endurance so that it can be stable at high voltages to ensure the normal running of the battery. For batteries with high-energy-density cathode materials, the CEI should also possess high mechanical strength to withstand a huge volume change of positive particles during charging and discharging processes [30, 31]. In addition, with the increasing studies on batteries in extreme conditions such as high-voltage batteries and high-temperature batteries, the CEI regulation becomes important unprecedentedly. The reported studies have proved that the CEI with optimized structures can effectively stabilize the electrolyte by inhibiting oxidation of the electrolyte at high voltages. Besides, the CEI after rational adjustment features a robust structure, thus endures the expansion stress of cathode particles during cycling [32]. In this section, we give a full discussion about the recent studies of CEI regulation for batteries.

2.3.2 Types and Modification Strategies

Li and coworkers reported on the study of (2-allylphenoxy) trimethylsilane (APTS) as additives to construct the CEI with high LiF content. The DFT calculations and NMR results demonstrated that APTS could strongly coordinate with HF generated from the hydrolysis of $PF6^-$. Besides, the higher oxidation tendency of APTS after coordinated with F was demonstrated by DFT calculations, and linear potential

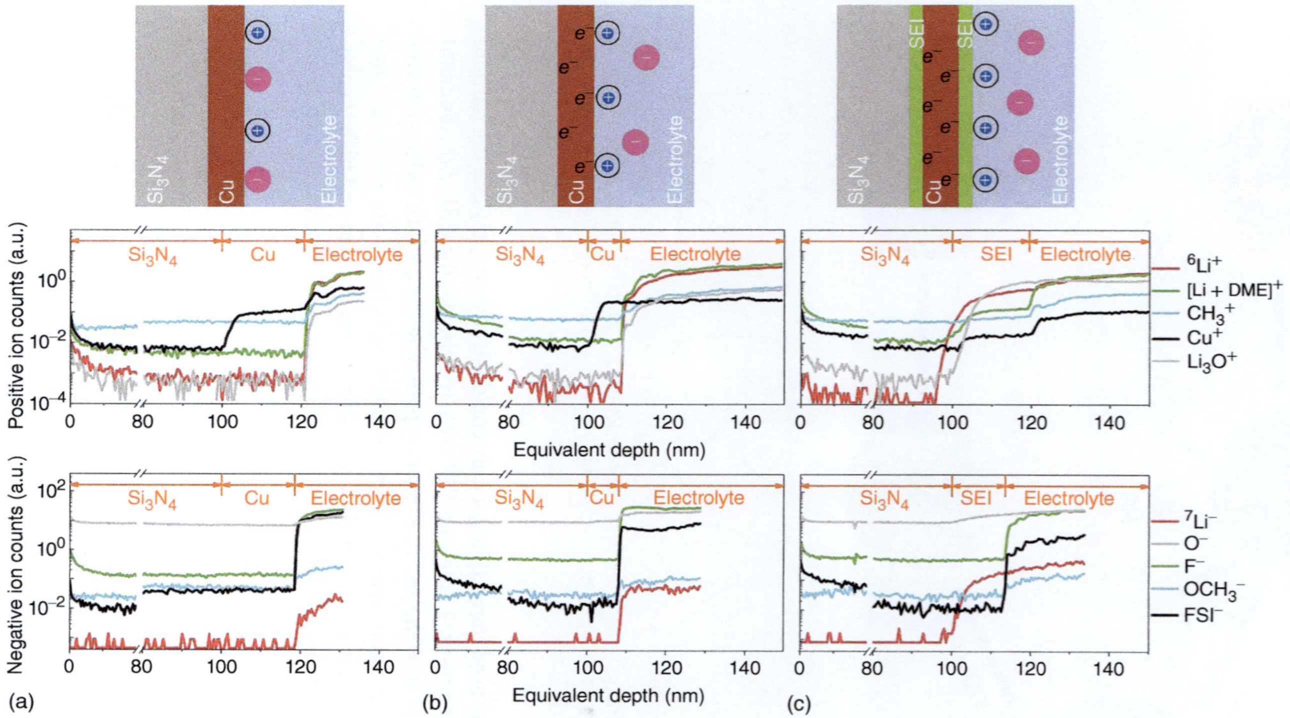

Figure 2.10 A schematic illustration of the formation of an SEI layer at the Cu anode surface with 1.0 M LiFSI in DME as the electrolyte and the corresponding positive ion and negative ion SIMS depth profiles. (a–c) Top rows: a fresh cell, OCP, both anions (FSI – stay on the Cu electrode surface (a); at 1.0 V, an electrical double layer forms and the electric force attracts solvated Li^+) and cations (solvated Li^+ ions), which leads to a better wettability of the electrolyte on the Cu surface (b); at 2.0 V, the inner SEI layer forms at the electrode/electrolyte interface, which includes the back side of the Cu electrode, because a porous Cu electrode is used (c); at 3.0 V, a small amount of Li metal may already be deposited on the Cu electrode surface. Source: Reproduced with permission Zhou et al. [29].

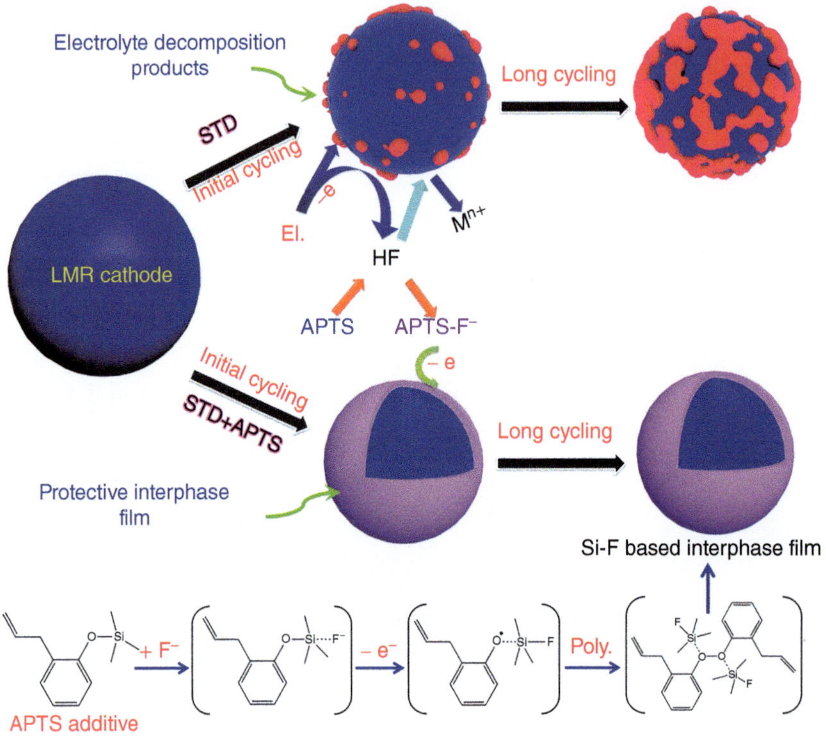

Figure 2.11 Schematic illustration of the construction of a protective interphase film on LMR through the transformation of F^- coordinating with APTS. Source: Reproduced with permission Ye et al. [23].

voltammograms proved that APTS not only could effectively eliminate the harmful HF but also could convert F^- into the CEI layer to generate the CEI with abundant LiF (Figure 2.11) [33]. Wang and coworkers reported on the 2% LiDFOB as additives into the fluorinated electrolyte for the stable cycle of battery with the cobalt-free $LiNiO_2$ cathode. The cobalt-free $LiNiO_2$ cathode features low cost and high theoretical energy density, but its practical application is hindered by continuous Ni dissolution and structural instability. By the addition of LiDFOB, a robust fluoride (F)- and boron (B)-rich CEI on the cathode was achieved (Figure 2.12). It has been demonstrated that the F-rich CEI displays low electronic conductivity and therefore can effectively reduce the interface resistance and prevent continuous decomposition of the electrolyte. Besides, the F- and B-rich CEI can also protect the $LiNiO_2$ cathode from continuous Ni dissolution. The time-of-flight secondary ion mass spectrometry (ToF-SIMS) shows that the CEI consists of Li and O components in abundance, while high F and B signals in the outer layer. The decreased C and O species demonstrated that the side reaction of the electrolyte with $LiNiO_2$ cathode was effectively suppressed (Figure 2.12) [34].

Guo and coworkers reported the in situ construction of amorphous CEI on a Ni-rich cathode with a hybrid electrolyte for stable Li metal batteries. Specifically,

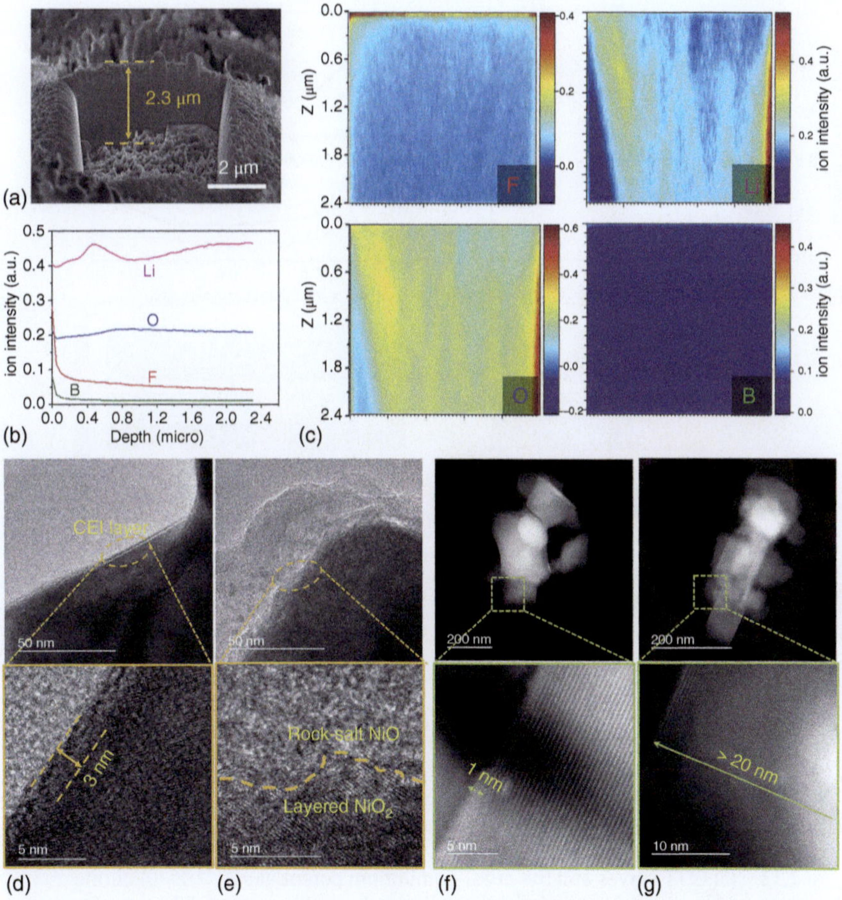

Figure 2.12 (a) Surface morphology of LNO particles after 100 cycles in F-262A for time-of-flight secondary ion mass spectrometry (ToF-SIMS) and the crater is sputtered by Ga^+ ion beam. (b) The corresponding lithium (Li), oxygen (O), fluorine (F), and boron (B) element distributions in the sputtered active particle. (c) The depth profiles of ToF-SIMS analysis for F, Li, O, and B elements on the cycled active particles. (d and e) High-resolution transmission electron microscopy (HRTEM) images of (d) LNO cathode particle after 150 cycles in F-262A electrolyte, covering with a thin CEI layer on surface (~3 nm); (e) LNO cathode particles after 70 cycles in E-baseline electrolyte – below images are enlarged views of the particles. (f and g) STEM images of (f) LNO cathode after 100 cycles in F-262A electrolyte with several layers of rock-salt NiO (~1 nm); (g) LNO cathode particles after 100 cycles in E-baseline with thick rock-salt NiO structure (>20 nm) – below images are enlarged views of particles. Source: Deng et al. [34]/with permission of Elsevier.

LiDFOB was used as the additive for the generation of CEI with an amorphous buffer layer ($Li_xBO_yF_z$) that can reinforce the interface of cathode/electrolyte to withstand high working voltage. The morphology characterization and ToF-SIMS show that the uniform CEI with amorphous spread on the surface of primary particles homogeneously, the $Li_xBO_yF_z$ protection layer displays the average thickness of 9 nm, which could contact with cathode well to realize the fast Li^+ diffusion

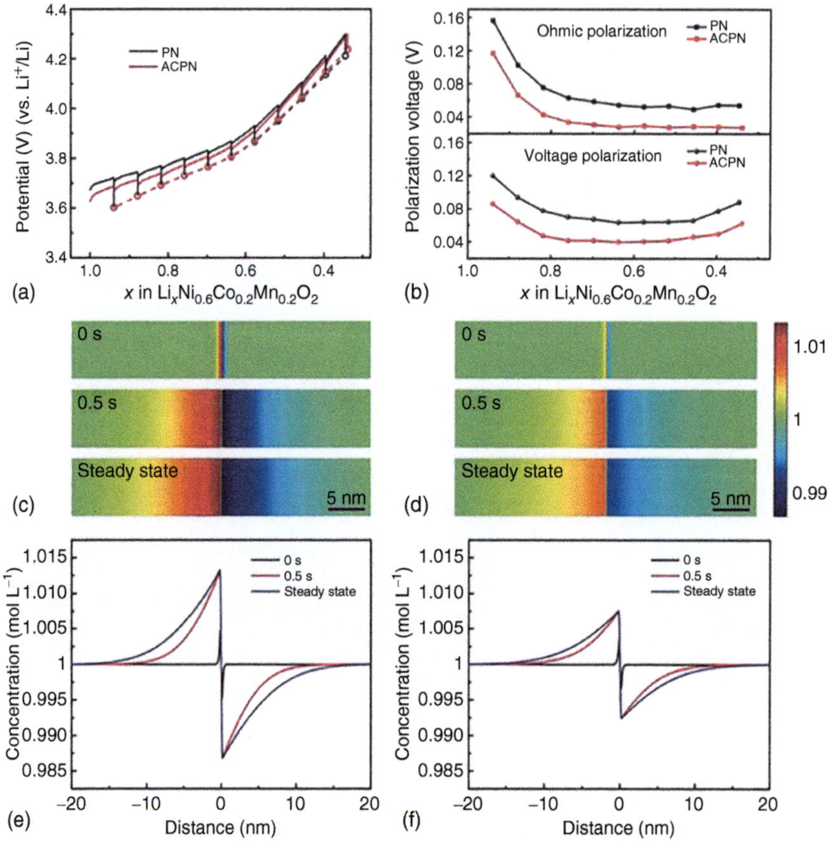

Figure 2.13 (a) GITT curves and the quasi-equilibrium potential vs. Li^+/Li (0) during charging and (b) corresponding polarization voltage derived from the GITT plots of two cathodes. (c–f) Li-ion concentration distribution and the corresponding curves extending opposite the cathode/electrolyte interface (middle boundary) at discharged state obtained from the COMSOL simulation for (c, e) PN and (d, f) ACPN, where the left side is cathode and the right side is the mhybrid solid/liquid electrolyte or CEI layer. Source: Liang et al. [5], figure 02 (p. 03)/with permission of John Wiley and sons.

on the interface. The COMSOL Multiphysics concentration field simulation was conducted to explore the interfacial kinetics of Li^+ on the interface (Figure 2.13). It was demonstrated that after the introduction of the additive, the potential drop on the interface was shifted from the steep potential drop to the gradual potential drop, which effectively reduced the diffusion impedance of Li^+ and facilitated the ion transportation on the interface [5].

Peng and coworkers proposed using lithium difluorophosphate ($LiPO_2F_2$) and fluoroethylene carbonate (FEC) as co-additives for conventional electrolytes to reduce the charge-transfer impedance and extend the cycle life of batteries with $LiNi_{1/3}Co_{1/3}Mn_{1/3}O_2$ cathodes (Figure 2.14). The calculated HOMO/LUMO energy levels of electrolyte components demonstrated that $LiPO_2F_2$ and FEC tend to be easily oxidized on the cathode to form the CEI on the cathode, which is consistent with the results of the linear scan voltammetry (LSV) curves. The reduplicated CVs of the battery using the electrolyte with additives show that with fast Li^+ diffusion kinetics, low polarization could be achieved. The much higher coulombic efficiency of electrolyte (99%) with additives than the baseline electrolyte (93.5%) again demonstrated the effectiveness of additives. By the EIS, they confirmed that the electrolyte with additives display much lower CEI film resistance and a much faster Li^+ diffusion coefficient. The morphologies of the cathode particles show that the CEI shifted from the aggregation of nonuniform particles to thin and uniform after the introduction of additives. XPS and TOF-SIMS demonstrated that the CEI featured a hybrid structure consisting of LiF, $LiCO_3$, and P–O species and such a structure is beneficial to avoid the decomposition of electrolyte and the structural degradation of the NMC cathode [35].

Kang and coworkers reported on the introduction of LiDFOB as the additive for the construction of hierarchical CEI for $LiNi_{0.8}Co_{0.1}Mn_{0.1}O_2$/Li metal batteries. The $LiNi_{0.8}Co_{0.1}Mn_{0.1}O_2$ (NCM811) cathode exhibits high energy density but with fast capacity degradation. By introducing LiDFOB, they found that the unique hierarchical CEI with LiF in the inner layer and organic species in the outer layer not only effectively prevented the side reaction of the Ni-rich cathode but also induced uniform deposition of Li^+. LiDFOB with two B-F bonds and an oxalic ring exhibited high ion conductivity and film-forming ability. The larger peak current density of CV curves and the slope of peak current density with a scan rate test in the battery with additives demonstrated that the battery with additives exhibited better kinetic properties (Figure 2.15). The EIS and TEM results also demonstrated that the uniform CEI with reduced Li^+ transfer resistance was successfully in situ generated on the surface of cathode particles homogeneously. The XPS characterization demonstrated that organic species such as $Li_xPO_yF_z$ and $Li_xBO_yF_z$ were mainly formed in the outer layer of the CEI, while the inner layer mainly consisted of LiF. Such a unique hybrid CEI can stay chemically stable during the cycles, thus can enhance the battery performance [36].

For a better understanding of the structural evolution of the CEI during cycling, Manthiram and coworkers used TOF-SIMS with XPS to comprehensively study the fragment-specific information of the CEI in the electrodes of high-Ni cathodes, Li-rich cathode, and high-voltage spinel cathode, respectively. The advanced TOF-SIMS characterization technique featuring submicron rastering capabilities and sub-nanometer depth resolution can provide in-depth details of the chemical components and the concentration of various species in the interface for these three cathode materials. They found that for the high-Ni cathode, the CEI consisted of abundant $LiCO_3$, which was derived from the high reactivity of the high-Ni cathode with air in the humid environment. For the spinel cathode, high content of C_2HO^- species was detected in the CEI during the early cycling stage, which suggests that the formation of CEI was dominated by the decomposition of $LiPF_6$. As for the Li-rich cathode, the polymeric carbon species were detected during the initial cycles. The corresponding results could also be observed on the mapping of the CEI with TOF-SI MS [37].

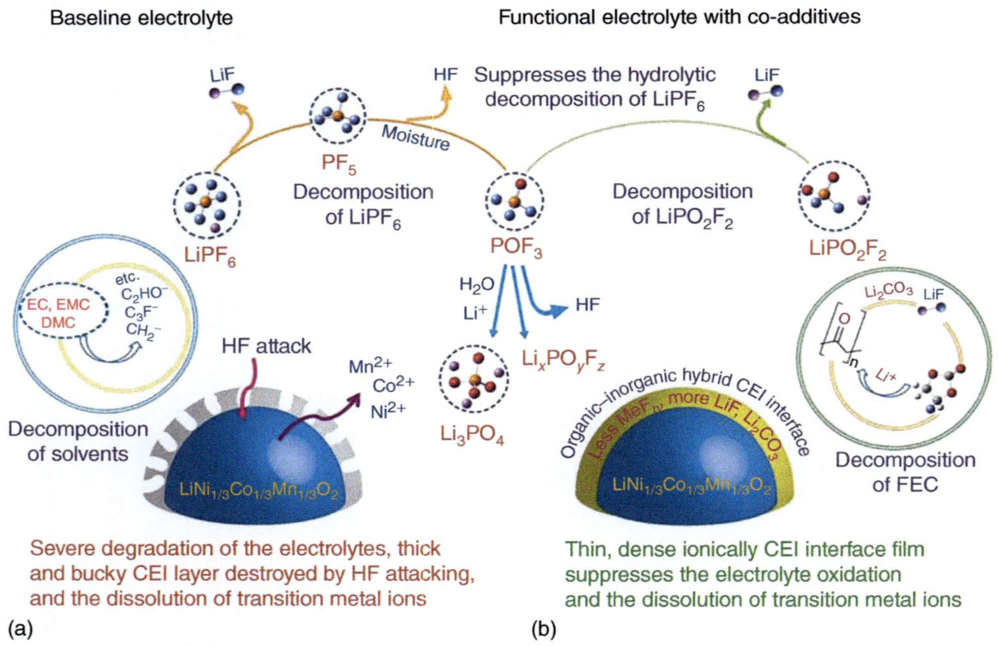

Figure 2.14 Schematic representation of the key effect posed by the use of $LiPO_2F_2$ (a) and FEC additives (b). Source: Reproduced with permission Chen et al. [35].

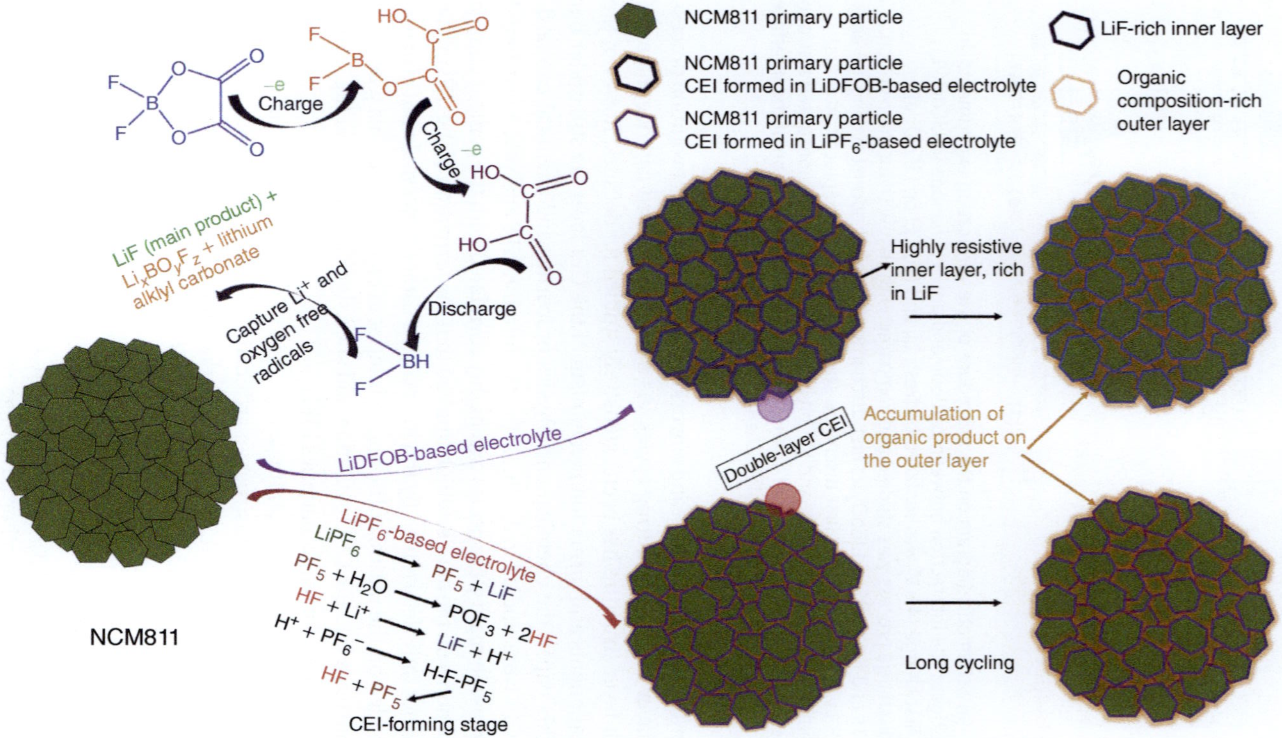

Figure 2.15 Scheme of CEI formation and revolution on NCM811 cathode using LiDFOB and $LiPF_6$ based electrolytes during cycling. Source: Reproduced with permission Mao et al. [36].

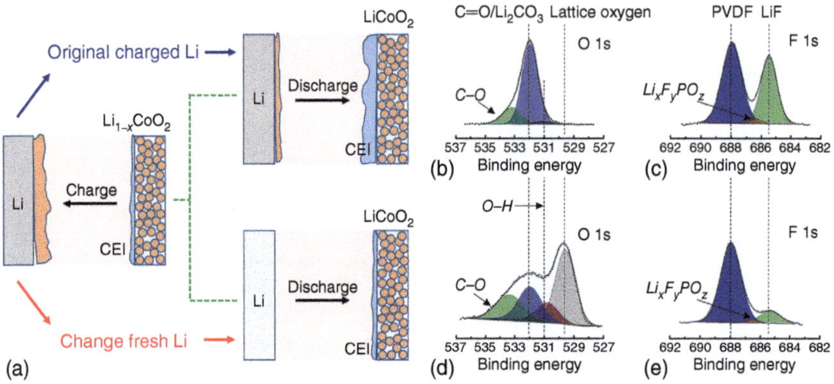

Figure 2.16 (a) Schematic diagram on the procedure of the verification experiments. In order to investigate the interaction between cathode and anode, Li anode in $LiCoO_2$/Li battery after first charge (4.6 V) was replaced by fresh Li, and the CEI on $LiCoO_2$ after subsequent discharge process was studied and compared with that without interruption. (b) and (c) O 1s and F 1s XPS of $LiCoO_2$ at first discharge state after normal cycling. (d) and (e) O 1s and F 1s XPS of $LiCoO_2$ at first discharge state with Li anode replaced after first charge. The amount change of Li_2CO_3 and LiF on cathode and anode demonstrated a strong correlation between $LiCoO_2$ cathode and Li anode on the dynamic evolution of the interface layer. Source: Reproduced with permission Zhang et al. [38].

From a similar starting point, Li and coworkers studied dynamic evolution process of the CEI layer on $LiCoO_2$ during cycles. As a commercialized cathode material, $LiCoO_2$ with layer structure has been extensively studied for decades; however, the detailed components and formation mechanism of the CEI layer are still under hot debate. In this paper, the chemical and physics properties and chemical compositions of the CEI were studied in depth. By the quantitative XPS analysis, it was found that the increase in main CEI species was highly correlated with the content change of SEI components on the anode, which implied that the formation of the CEI on the cathode was not only determined by the interface of cathode/electrode but also was highly depended on the nature of the anode interface (Figure 2.16). This work proposed new insights into the CEI regulation and shed significant light on the further design of electrolytes for stable batteries [38].

Besides the above-discussed strategies of electrolyte regulation for the CEI, another unique method is to construct an artificial CEI layer on the cathode directly by an ex situ or in situ method. Cho and coworkers created an artificial CEI on the cathode by a simple method involving a reaction of residual lithium compounds with cobalt phosphates. They found that the original artificial CEI on the cathode reacted with the by-products of batteries and grew along the grain boundaries to form the uniform CEI. The as-formed CEI exhibited high thermal stability and maintained structural integrity, thus ensured the long cycling performance of the battery (Figure 2.17). Another advantage of this method is that the undesired residual lithium compounds such as LiOH and Li_2CO_3 were eliminated without additional treatments; the as-formed CEI also eliminated the harmful species such as HF and H_2O [39].

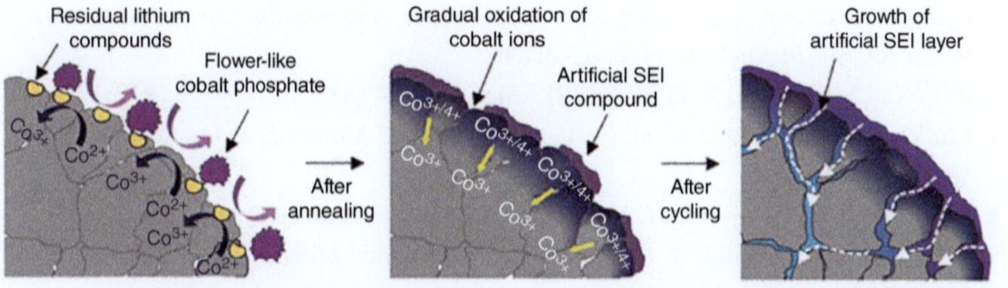

Figure 2.17 Schematic of the ASL–NCA preparation, showing the cobalt migration on the surface and the growth of the artificial SEI layer during battery cycling. Source: Reproduced with permission Kim et al. [39].

In general, the electrode–electrolyte interphase on both anode and cathode acts as the protection layer to prevent side reactions of electrodes with electrolyte. For the anode, the SEI with ideal properties also greatly suppresses the growth of lithium dendrites and regulates the Li dissolution kinetics. For the cathode, the dense and robust CEI effectively suppresses the dissolution of transition metal ions and can absorb harmful by-products such as HF in the electrolyte. In this chapter, the method of boosting the performance of Li metal batteries based on the adjustment of both SEI and CEI was briefly reviewed. The regulation of SEI by the adjustment of electrolyte based on various mechanisms was summarized as follows: (i) Design of compatible electrolytes with the selected electrodes in different systems to ensure the normal running of battery; (ii) regulation of the Li solvation structure by anion regulation to influence the features of SEI; (iii) introduction of additives to generate the SEI with various structures and components such as SEI rich in inorganic species and hybrid SEI with LiF–organic bilayer structure. As for the CEI, similar guidelines can be used. Besides, some unique insights such as the CEI can also be influenced by the nature of SEI have also been proposed. In addition, employing an artificial CEI layer on the cathode directly was also developed. In future studies, advanced characterization techniques should be applied to study the interdependent factors that influence the SEI/CEI together. The functionality of different component species in SEI/CEI should be clearly identified to guide the design of SEI/CEI with ideal properties.

References

1 Ming, J., Cao, Z., Wu, Y. et al. (2019). *ACS Energy Letters* 4: 2613–2622.

2 Wu, B., Wang, S., Lochala, J. et al. (2018). *Energy & Environmental Science* 11: 1803–1810.

3 Zhang, W., Shen, Z., Li, S. et al. (2020). *Advanced Functional Materials*: 2003800. https://doi.org/10.1002/adfm.202003800.

4 Song, H.J., Jang, S.H., Ahn, J. et al. (2019). *Journal of Power Sources* 416: 1–8.

5 Liang, J.Y., Zhang, X.D., Zeng, X.X. et al. (2020). *Angewandte Chemie (International Edition in English)* 59: 6585–6589.
6 Fan, X., Chen, L., Ji, X. et al. (2018). *Chem* 4: 174–185.
7 Lu, W., Zhang, J., Xu, J. et al. (2017). *ACS Applied Materials & Interfaces* 9: 19313–19318.
8 Liu, W., Li, J., Li, W. et al. (2020). *Nature Communications* 11: 3629.
9 Li, Q., Wang, Y., Wang, X. et al. (2020). *ACS Applied Materials & Interfaces* 12: 2319–2326.
10 Chae, B.-J. and Yim, T. (2017). *Journal of Power Sources* 360: 480–487.
11 Zhang, X.Q., Chen, X., Cheng, X.B. et al. (2018). *Angewandte Chemie (International Edition in English)* 57: 5301–5305.
12 Lee, J.I., Song, G., Cho, S. et al. (2020). *Batteries & Supercaps* 3: 828–859.
13 Hou, Z., Zhang, J., Wang, W. et al. (2020). *Journal of Energy Chemistry* 45: 7–17.
14 Zhang, X.Q., Li, T., Li, B.Q. et al. (2020). *Angewandte Chemie (International Edition in English)* 59: 3252–3257.
15 Yang, H., Li, J., Sun, Z. et al. (2020). *Energy Storage Materials* 30: 113–129.
16 Wang, R., Cui, W., Chu, F., and Wu, F. (2020). *Journal of Energy Chemistry* 48: 145–159.
17 Peled, E. and Menkin, S. (2017). *Journal of the Electrochemical Society* 164: A1703–A1719.
18 Wang, A., Kadam, S., Li, H. et al. (2018). *Computational Materials* 4 (1): 1–26.
19 Xu, R., Yan, C., Xiao, Y. et al. (2020). *Energy Storage Materials* 28: 401–406.
20 Zhang, X.-Q., Chen, X., Hou, L.-P. et al. (2019). *ACS Energy Letters* 4: 411–416.
21 Fu, J., Ji, X., Chen, J. et al. (2020). *Angewandte Chemie (International Edition in English)* 59: 22194–22201.
22 Wang, Z., Qi, F., Yin, L. et al. (2020). *Advanced Energy Materials* 10: 1903843.
23 Meng, J., Chu, F., Hu, J., and Li, C. (2019). *Advanced Functional Materials* 29: 1902220.
24 Dai, H., Gu, X., Dong, J. et al. (2020). *Nature Communications* 11: 643.
25 Chen, J., Fan, X., Li, Q. et al. (2020). *Nature Energy* 5: 386–397.
26 Tan, S.J., Yue, J., Hu, X.C. et al. (2019). *Angewandte Chemie (International Edition in English)* 58: 7802–7807.
27 Qi, S., Wang, H., He, J. et al. (2020). *Scientific Bulletin* https://doi.org/10.1016/j.scib.2020.1009.1018.
28 Wang, H., He, J., Liu, J. et al. (2020). *Advanced Functional Materials*: 2002578. https://doi.org/10.1002/adfm.202002578.
29 Zhou, Y., Su, M., Yu, X. et al. (2020). *Nature Nanotechnology* 15: 224–230.
30 Huang, Y., Cao, S., Xie, X. et al. (2020). *ACS Applied Materials & Interfaces* 12: 19483–19494.
31 Yang, W., Xiang, W., Chen, Y.X. et al. (2020). *ACS Applied Materials & Interfaces* 12: 10240–10251.
32 (a) Song, S.H., Cho, M., Park, I. et al. (2020). *Advanced Energy Materials* 10: 2000521. (b) Li, Q., Zhuang, W., Li, Z. et al. (2020). *ChemElectroChem* 7: 998–1006.

33 Ye, C., Tu, W., Yin, L. et al. (2018). *Journal of Materials Chemistry A* 6: 17642–17652.

34 Deng, T., Fan, X., Cao, L. et al. (2019). *Joule* 3: 2550–2564.

35 Chen, Y., Zhao, W., Zhang, Q. et al. (2020). *Advanced Functional Materials* 30: 2000396.

36 Mao, M., Huang, B., Li, Q. et al. (2020). *Nano Energy* 78: 105282.

37 Erickson, E.M., Li, W., Dolocan, A., and Manthiram, A. (2020). *ACS Applied Materials & Interfaces* 12: 16451–16461.

38 Zhang, J.-N., Li, Q., Wang, Y. et al. (2018). *Energy Storage Materials* 14: 1–7.

39 Kim, J., Lee, J., Ma, H. et al. (2018). *Advanced Materials* 30: 1704309.

3

Safe Electrolytes

3.1 Introduction

Since the commercialization of lithium-ion batteries (LIBs), they are widely used in portable electronic devices such as mobile phones, computers [1, 2]. With the further development of the electric vehicle industry, the demand for high-energy batteries has been increasing [3–5]. However, safety issues Li power batteries are not yet resolved. To improve the safety of Li power batteries is a basic guarantee for their future development and application. The safety problem of Li power batteries is mainly related to electrolytes [6–8]. Generally, electrolytes are nonaqueous solvents which are flammable organic compounds. For example, ethylene carbonate (EC), dimethyl carbonate (DMC), propylene carbonate (PC), diethyl carbonate (DEC), and 1,2-dimethoxyethane (DME) [9–11]. Under abuse conditions such as short-circuiting, overcharging, high temperature, and so on, combustion of Li metal and electrolyte will result in thermal runaway, thereby causing fire and explosion in Li power batteries [12, 13]. Therefore, it is important to develop flame-retardant and even nonflammable electrolytes to improve the safety of Li batteries.

In recent decades, extensive research has been done to reduce flammability of electrolytes [14, 15]. Solid-state electrolytes have low flammability; however, they suffer from high impedance and low Li ionic conductivity [16, 17]. Therefore, adding flame-retardant additives or nonflammable solvents into conventional electrolytes can reduce or even eliminate their flammability [18, 19]. Extensive flame-retardant and nonflammable electrolytes have been developed that can significantly reduce fire and explosion risks of LIBs [20, 21]. With rapid growth of energy demand, the design and development of high-energy-density energy storage equipment becomes very urgent. Li metal batteries with ultrahigh theoretical capacity (3860 $mAh\,g^{-1}$) and lowest reduction potential (−3.04 V vs. RHE [reversible hydrogen electrode]) have received attention from scientists [22, 23]. Li metal batteries (LMBs) with Li metal anodes have energy density several times higher than LIBs. Compared to graphite, Li metal anodes can achieve higher capacity [24]. However, Li can easily form dendrites than graphite anodes. Overgrowth of Li dendrites will lead to excessive consumption of electrolyte and Li metal even piercing separators, resulting in battery short circuit, which exacerbates potential safety issues [25–27]. Therefore, more studies are required to ship high-safety LMBs [28–31]. In this

Liquid Electrolyte Chemistry for Lithium Metal Batteries: Design, Mechanisms, Strategies, First Edition. Jianmin Ma.

chapter, we review the flame-retardant mechanism and design principles and hope to provide guidelines for the design of efficient flame-retardant electrolytes.

3.2 Flame-Retardant Mechanism

In a combustion process, fuel, combustion aids, and heat are necessary components to keep the fire until at least one of those gets exhausted. So flame retardants are divided into two types: gas-phase and condensed-phase [32]. In the combustion of liquid electrolytes, solvents will produce hydrogen radicals in the flame (Eq. (3.1)). Thermal decomposition of cathode materials will produce oxygen. Then, oxygen will react with hydrogen radicals to produce hydroxide and oxygen radicals (Eq. (3.2)). The reduction of the electrolytes will produce hydrogen (Eq. (3.3)) and further react with oxygen and hydroxide radicals to generate more radicals, thus leading to flame or even explosion (Eqs. (3.4) and (3.5)). So the key to suppress fire caused by the liquid electrolytes is to eliminate hydrogen radicals in the electrolytes.

$$RH \rightarrow R^{\cdot} + H^{\cdot} \tag{3.1}$$

$$O_2 + H^{\cdot} \rightarrow O^{\cdot} + HO^{\cdot} \tag{3.2}$$

$$RH \rightarrow R^{\cdot} + H_2 \tag{3.3}$$

$$HO^{\cdot} + H_2 \rightarrow H_2O + H^{\cdot} \tag{3.4}$$

$$O^{\cdot} + H_2 \rightarrow HO^{\cdot} + H^{\cdot} \tag{3.5}$$

Phosphorous- and halogen-containing chemical compounds are the commonly used flame retardants. For example, phosphorous evaporates and reaches the flame when the liquid electrolytes burn. Then, phosphorous is decomposed into P-containing radicals. These radicals can scavenge the active agents of hydrogen radicals; thus the combustion reaction is suppressed (Eqs. (3.6) and (3.7)). Similarly, Si-, N-, and F-containing additives follow the same flame-retardant principle [33, 34].

$$\text{Phosphorous} \rightarrow [P]^{\cdot} \tag{3.6}$$

$$[P]^{\cdot} + H^{\cdot} \rightarrow [P]H \tag{3.7}$$

3.3 Flame-Retardant Electrolytes

In the flame-retardant electrolyte field, the most commonly used index is self-extinguishing time, which is the continuous burning time of an ignited electrolyte sample. As suggested by Xu et al. flame-retardant electrolytes are divided into three kinds: "flammable electrolytes" with the self extinguishing time (SET) greater than 20 seconds g^{-1}; "flame-retarded electrolytes" with the SET between 6 and 20 seconds g^{-1}; and "nonflammable electrolytes" with the SET lower than 6 seconds g^{-1} [35].

Various phosphates have been added as additives or solvents to conventional electrolytes due to their better flame retardancy. Due to better flame retardancy, low-viscosity trimethyl phosphate (TMP) has been used as additives to improve the safety of lithium cells. For example, about 5, 10, and 20 wt% TMP as additives

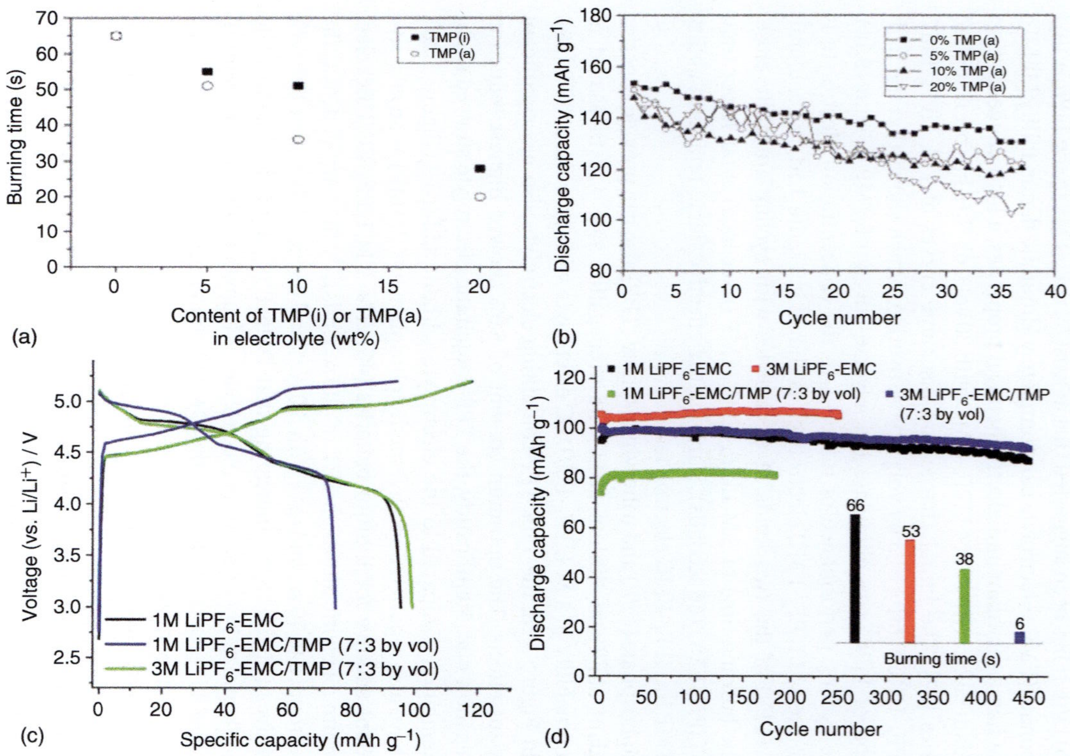

Figure 3.1 (a) The burning time and (b) cycling performance of different contents of TMP in electrolytes. Source: Reproduced with permission Yao et al. [36]. (c) Initial charge–discharge curves of graphite electrodes in 1 M $LiPF_6$-EMC and 1 and 3 M $LiPF_6$-EMC/TMP (7 : 3 by vol). (d) The burning time and charge–discharge curves of 1 and 3 M $LiPF_6$-EMC and 1 and 3 M $LiPF_6$-EMC/TMP (7 : 3 by vol). Source: Reproduced with permission Zhang et al. [37].

have been added into the baseline electrolyte of 1 M $LiPF_6$ in the ethylene carbonate/diethyl carbonate (EC/DEC) electrolyte. The electrolyte's flammability is shown in Figure 3.1a; the flammability of the electrolyte decreases with the addition of TMP into the electrolyte. So TMP can be used as an effective flame-retarding additive. However, the electrochemical performance of electrolytes containing flame retardants is poor; the TMP-containing cells showed rapid capacity loss (Figure 3.1b) [36]. It has been suggested that at least 62 vol% TMP in the $LiPF_6$-ethyl methyl carbonate (EMC) electrolyte can completely eliminate flammability [38]. To optimize the flame-retardant and electrochemical properties, a high-concentration flame-retardant electrolyte of 3 M $LiPF_6$-EMC with 30 vol% of TMP has been applied in Li batteries [37]. The initial charge and discharge curves of 1 M $LiPF_6$-EMC and 3 M $LiPF_6$-EMC/TMP (7 : 3 by vol) almost overlap, which shows that the high concentration of $LiPF_6$ can eliminate the adverse effects of TMP (Figure 3.1c). Further, the cycling performance of full cells in the concentration electrolyte is compared to the 1 M $LiPF_6$-EMC electrolyte. As shown in Figure 3.1d, the coulomb efficiencies (CEs) of the two kinds of electrolytes with cycling are quite close. However, the burning time of 3 M $LiPF_6$-EMC/TMP (6 seconds) is far lower than that of 1 M $LiPF_6$-EMC (66 seconds), implying the effective flame-retardant ability of 30 vol% TMP in $LiPF_6$-EMC electrolyte.

Triphenylphosphate (TPP) and tributylphosphate (TBP) also can provide superior flame-retardant ability in lithium cells. The propagation rate was measured to evaluate flammability in 1 M $LiPF_6$-EC/DEC electrolyte with or without TPP [39]. As shown in Figure 3.2a, the 1.0% TPP-containing electrolyte also has significant flame-retardant effects. The optimum content of 5.0% showed highest reduction in the propagation rate. Significantly, the TPP-containing electrolyte showed slight capacity attenuation compared to the electrolyte without additives (Figure 3.2b). Chen et al. have also reported dimethyl methylphosphonate (DMMP) as an effective flame-retardant additive for lithium batteries [40]. The 1 M $LiPF_6$-EC/DEC electrolyte is flammable with ultrahigh SET of 140 seconds g^{-1}. In sharp contrast, the SET greatly dropped to 70 seconds g^{-1} with 5.0 wt% and 0 seconds g^{-1} with 10.0 wt% DMMP additives (Figure 3.2c), indicating that DMMP is an efficient additive to suppress the flammability of baseline electrolytes. Meanwhile, the electrochemical performance of Li||$LiCoO_2$ cells with the DMMP-containing electrolyte was only slightly decreased compared to the baseline electrolyte (Figure 3.2d). Therefore, DMMP is also a potential flame retardant to enhance the safety of lithium batteries.

Recently, high-concentration electrolytes (HCEs) are attracting more attention due to the ultrahigh salt–solvent ratio and unique solvation structure [41, 42]. Such electrolytes can form high-stable solid electrolyte interphase (SEI) and cathode electrolyte interphase (CEI) to protect Li metal anodes and cathodes. It has also been found that HCEs have excellent flame-retardant ability [43]. As shown in Figure 3.3a,b, with the increase in Li-salt concentration from 1 to 6.5 M, the free fluoroethylene carbonate (FEC) gradually turned into coordinated FEC. With the increasing ratio of LiTFSI to FEC, the formation of solvate complexes from contact ion pairs to the aggregates indicates that more FEC molecule are

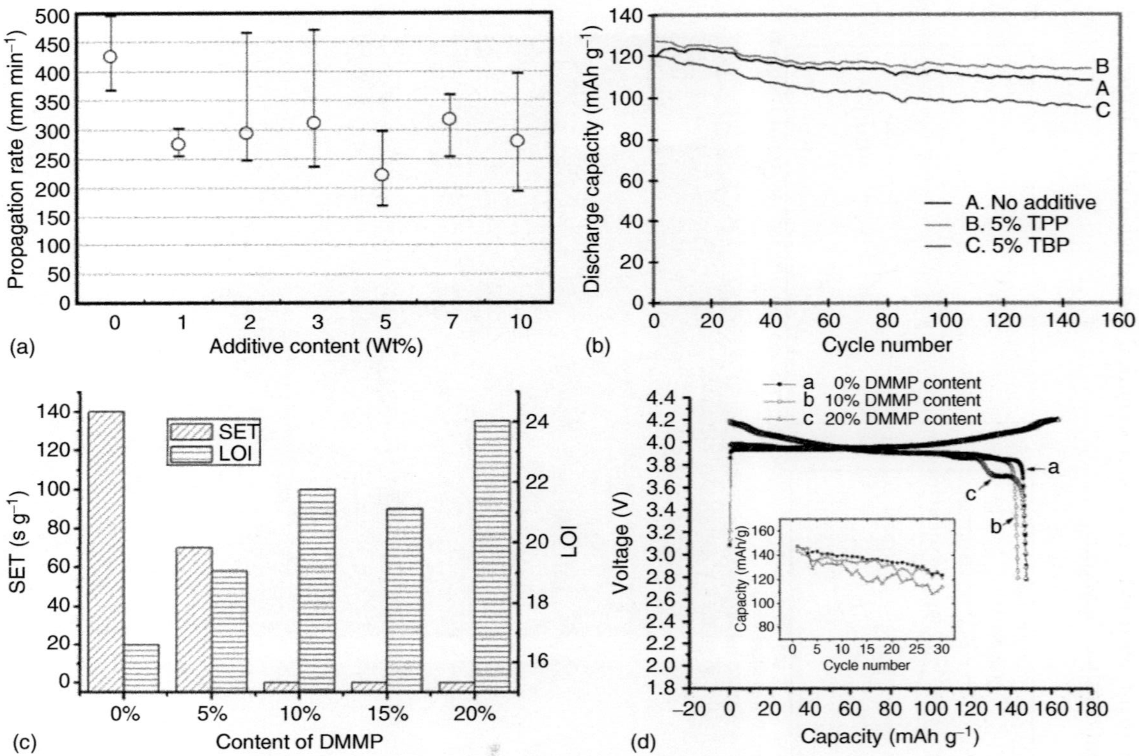

Figure 3.2 (a) The propagation rate of different contents of TPP-containing electrolyte. (b) The cycling performance of baseline, 5.0% TPP and 5.0 TBP-containing electrolytes. Source: Reproduced with permission Hyung et al. [39]. (c) The flammability test of SET and (d) electrochemical performance of Li||$LiCoO_2$ cells for different percentages of DMMP-containing electrolyte. Source: Reproduced with permission Xiang et al. [40].

coordinated with Li^+.. The glass fiber soaked in all electrolytes was ignited to evaluate the flame retardancy. In dilute solutions (Figure 3.3c,d), they are easily ignited, suggesting these two kinds of electrolytes are highly flammable. However, the 6.5 M LiTFSI/FEC electrolyte showed excellent flame retardancy even upon exposure to fire (Figure 3.3e). So the high-concentration LiTFSI/FEC electrolyte is a flame-retardant electrolyte. The electrochemical performance of the battery can also be significantly enhanced.

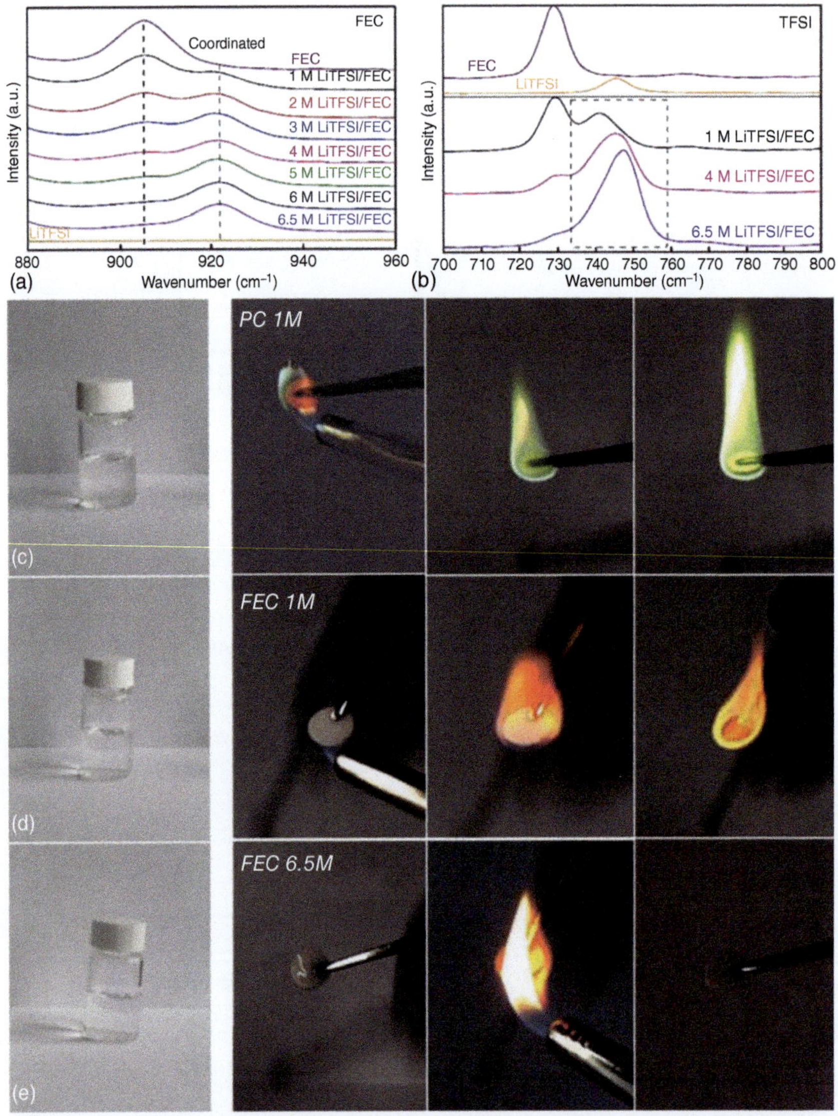

Figure 3.3 Raman spectra of LiTFSI/FEC mixtures (a) from 880 to 960 cm^{-1}, (b) 700 to 800 cm^{-1}. Flame-retardant pictures of (c) 1 M LiDFOB/PC, (d) 1 M LiTFSI/FEC, and (e) 6.5 M LiTFSI/FEC. Source: Yu et al. [43] / with permission of Elsevier.

3.4 Nonflammable Electrolytes

In general, the flame-retardant electrolyte is the result of compromise under the electrochemical and flame retardant properties. With the advancements in electrolyte engineering, many methods have been proposed using electrolyte additives and cosolvent methods, which can achieve high electrochemical performance with nonflammable electrolytes [44].

Triethyl phosphate (TEP) has good solvation of Li salts, low viscosity, high compatibility with carbonate electrolytes, and high flame retardancy [45]. It can be completely nonflammable as long as the content of TEP is higher than 14.0 vol% in carbonate electrolytes. However, TEP suffers from serious decomposition on the Li metal or graphite anode. So reducing the content of TEP and introducing film-forming additives are the best ways to adjust Li-battery performance. For example, Sun et al. have reported a nonflammable electrolyte consisting of 1 M lithium bis(trifluoromethanesulfonyl)imide (LiTFSI)-TEP with film-forming additives of vinylene carbonate (VC) and FEC. The carbonate electrolyte is flammable, whereas TEP electrolytes are nonflammable (Figure 3.4a,b). With the addition of film-forming additives, the long-term stability and CE are obviously enhanced due to the stable SEI film formed from VC and FEC [46].

In 2008, dimethyl(2-methoxyethoxy)methylphosphonate (DMMEMP) was first synthesized and used as a high safe solvent for LIBs by Li et al. [47]. As DMMEMP has low viscosity, high boiling point, and high dielectric constant, it is also a suitable flame-retardant electrolyte solvent. 1 M LiTFSI-DMMEMP is nonflammable having a wide electrochemical window at room temperature. Therefore, the $LiFePO_4||Li$ cell with LiTFSI-DMMEMP electrolyte delivers high capacity retention and CE.

As mentioned earlier, TEP is also a solvent that is not resistant to high voltage. To improve the stability under high-voltage conditions, fluorinated TEP of tris(2,2,2-trifluoroethyl)phosphate (TFEP) is a suitable solvent, which not only has

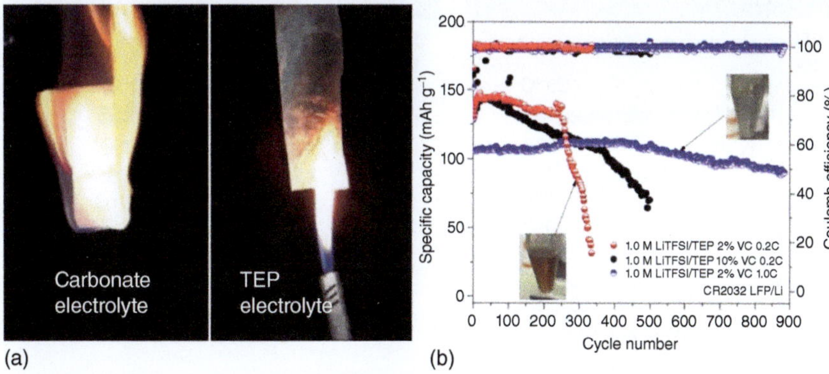

Figure 3.4 (a) Combustion tests of 1.0 M $LiPF_6$ EC/DEC (1 : 1) and 1.0 M LiTFSI/TEP 2.0% VC electrolytes. (b) Cycle performance of LFP||Li cells with 1.0 M LiTFSI/TEP 2.0% VC, 1.0 M LiTFSI/TEP 10.0% VC and 1.0 M LiTFSI/TEP 2.0% FEC electrolytes. Source: Jiang et al. [46] / with permission of American Chemical Society.

nonflammability and good compatibility with conventional electrolytes but also has an operation voltage higher than 4.5 V [11, 48]. Therefore, TFEP is considered an effective flame-retardant component for safe Li batteries [49]. Okuda et al. have reported an TFEP-based electrolyte with high nonflammability for lithium batteries [50]. TFEP possesses excellent self-extinguishing properties and is also an additive to form stable protective films to protect high-voltage cathodes. Ignition tests are carried out using separators saturated with different electrolytes. As shown in Figure 3.5a–d, the TFEP-containing electrolyte does not burn at all. The measurements of graphite||Li and $LiNi_{0.8}Co_{0.15}Al_{0.05}O_2$||Li cells show that the TFEP-containing electrolyte also possessed excellent stability. The capacity has almost no attenuation after 100 cycles test at room temperature condition (Figure 3.5e,f).

All fluorinated solvents such as fluoroethylene carbonate are nonflammable but often show a limited performance in LIBs and LMBs [51–53]. To improve the

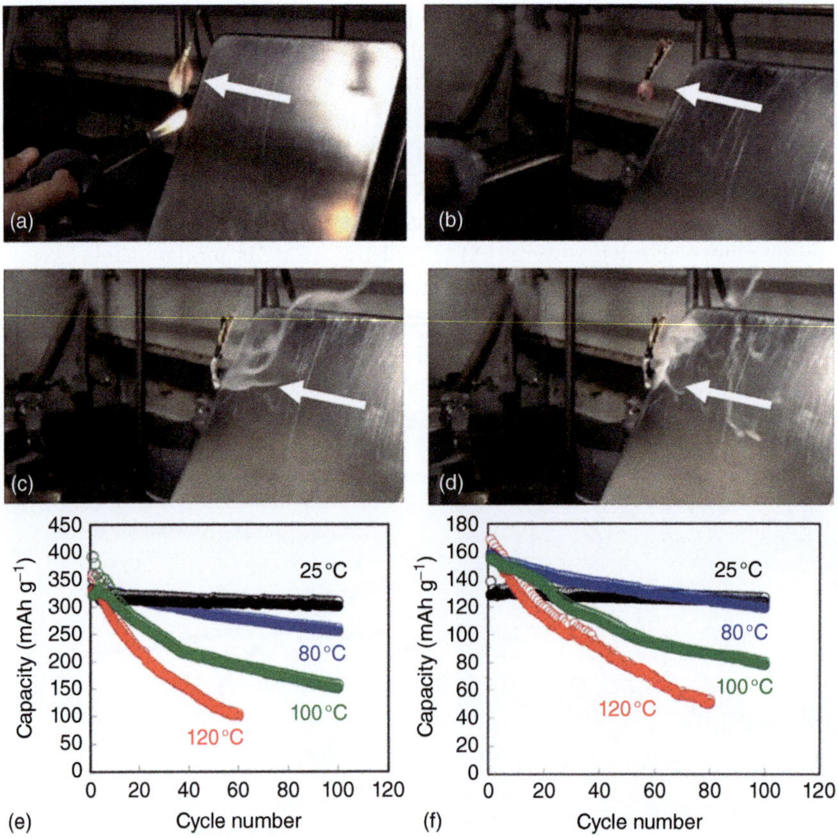

Figure 3.5 The ignition tests using (a) 1 M $LiPF_6$-EC/EMC, (b) super-concentrated electrolytes of [PC]/[LiFSA] = 2, (c) [TMP]/[LiFSA] = 2, and (d) [TFEP]/[LiFSA] = 2. Cycle performance of (e) graphite||Li and (f) $LiNi_{0.8}Co_{0.15}Al_{0.05}O_2$||Li cells with the [TFEP]/[LiFSA] = 2 electrolyte from 25 to 120 °C. Source: Shiga et al. [50] / with permission of Royal Society of Chemistry.

performance, the compounds of conventional electrolytes are optimized [54]. As shown in Figure 3.6, the EC/EMC-based electrolyte is flammable with an SET of 60 seconds g^{-1} (Figure 3.6a), whereas the SET of EC/FEMC-based electrolytes is found to be zero, indicating that these do not catch fire at all (Figure 3.6b,c). Significantly, the nonflammable EC/FEMC-based electrolyte exhibited high capacity compared to EC/FEMC-based electrolyte because the FEMC aids in formation of more stable interphase (Figure 3.7a,b).

In addition to P-, F-containing flame retardants, N-containing species such as nitriles possess the flame-retardant ability. Compared to carbonate solvents, nitriles with wide electrochemical window and high thermal stability have attracted huge attention for high-voltage and safety batteries. However, past studies have shown that nitrile-containing electrolytes often display a poor long-term stability [9, 55]. This is because nitrile electrolytes cannot form stable CEI or SEI. Therefore, it is an urgent problem to improve the stability of SEI and realize the application of nitriles as electrolytes with high safety and high voltage. Chen et al. have reported a succinonitrile (SN)-based dual-anion electrolyte as a safe electrolyte for LMBs [56]. For this electrolyte, the SN was selected as solvent and LiDFOB and LiTFSI as Li salts, which achieved superior stability of the interfacial phase toward LMBs. The flammability test showed that the liquid D-DES could be ignited by a lighter (Figure 3.8a). In contrast, the 1 M $LiPF_6$ EC/DMC electrolyte is easily ignited (Figure 3.8b). Surprisingly, such a electrolyte with D-DES showed excellent long-term performance in high-voltage $LiCoO_2$-based LMBs. As shown in Figure 3.8c,d, ultrahigh capacity retention and cycling stability of about 80% for 1000 cycles and 77% for 750 cycles were achieved with the voltages of 3.0–4.5 and 3.0–4.7, respectively. Due to high nonflammability and excellent interface stability of both the anode and cathode, safety accidents can be completely avoided. This strategy with dual-anion and deep eutectic solution showed high flame-retardant ability and superior electrochemical performance, which will greatly boost the practical applications of safe LMBs.

It is difficult to develop new P-, F-, and N-containing flame retardants. By contrast, modifying the traditional carbonate electrolytes is much simpler [57, 58]. Yang et al. have reported a tri-salt of $LiPF_6$, LiTFSI, and LiBOB in EC/PC electrolyte, which is nonflammable [59]. In this system, the EC and PC have high flash point and high dielectric constant combined with thermally stable salts of LiTFSI and LiBOB. In this strategy, the tri-salt electrolyte is not ignitable by fire. As shown in Figure 3.9a–c, the tri-salt in EC/PC electrolyte is nonflammable even upon exposure to fire (Figure 3.9a). The tri-salt in EC/DEC electrolyte is flammable (Figure 3.9c), but the combustion degree is not as high as $LiPF_6$ in EC/DEC electrolyte (Figure 3.9b), indicating that the synergistic effect of LiTFSI, LiBOB and PC is the necessary conditions for nonflammable. Meanwhile, a homogeneous and stable SEI has formed in the tri-salt electrolyte, thus displaying better electrochemical performance in NCM||Li cells.

HCEs and localized high-concentration electrolytes (LHCEs) are used to improve the stability of SEI to boost the long-term performance in LMBs [60–63]. Numerous studies have shown that salt-concentrated electrolytes can resolve the incompatibility between the flame-retardant ability and electrochemical performance of

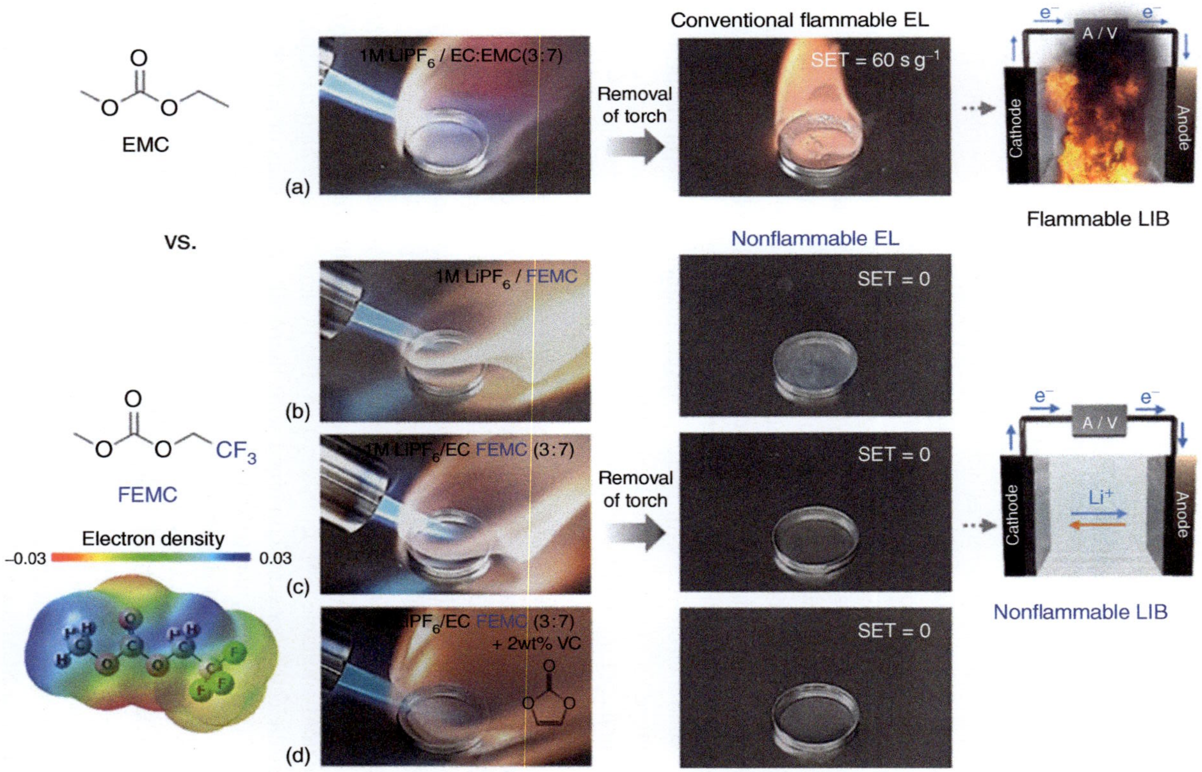

Figure 3.6 Self-extinguishing time test for (a) conventional electrolyte, (b) 1 M $LiPF_6$/FEMC, and designed EC/FEMC-based electrolyte (c) without and (d) with 2.0 wt% VC additive. Source: Chung et al. [54] / with permission of American Chemical Society.

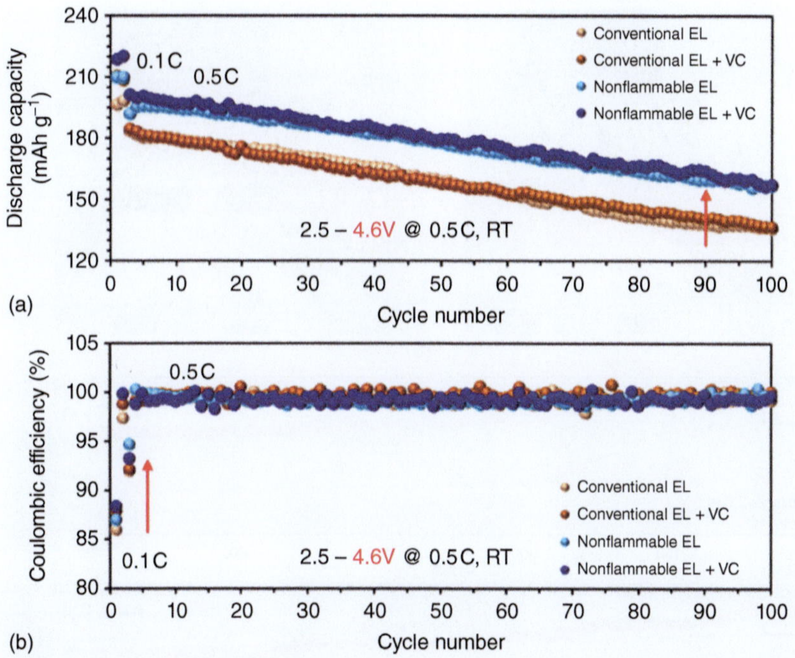

Figure 3.7 Cycling performance (a) and CE (b) of 1 M $LiPF_6$-EC/EMC without or with 2.0 wt% VC and 1 M $LiPF_6$-EC/FEMC without or with 2.0 wt% VC. Source: Reproduced with permission Chung et al. [54].

phosphate solvents. Because these solvents are not suitable for passivating graphite and Li metal anodes, electrochemical performance is generally poor (Figure 3.10) [64]. In HCEs, a robust salt-derived inorganic SEI is formed on the anode and high nonflammability can be guaranteed.

Xiang et al. developed a concentrated phosphate-based electrolyte consisting of 5 M LiFSI in TMP. This electrolyte is nonflammable compared to 1 M $LiPF_6$-EC/DEC electrolyte (Figure 3.11a,b) [65]. Since the high-concentration 5 M LiFSI/TMP electrolyte has good compatibility with the graphite anode, its electrochemical performance in graphite||NCM cell is excellent. In Figure 3.11c, the cell with 5 M LiFSI/TMP electrolyte had an initial capacity of 142.2 mAh g^{-1} and initial CE of 81.0%. After 100 cycles, the capacity and capacity retention ratio of the cell were 109.4 mAh g^{-1} and 84%, respectively. Chen et al. also reported a HCE of 4 M LiFSI in TMP [66]. The solvation structure of the electrolytes was regulated, which achieved nonflammability and suppressed Li dendrite growth. The symmetric Li cells with 4 M LiFSI/TMP and 6 M LiFSI/TMP electrolytes showed stable Li plating/stripping behaviors over 600 hours at a current density of 0.1 mA cm^{-2} (Figure 3.12a) and the cell with 4 M LiFSI/TMP electrolyte displayed the most stable cycling performance for 1000 hours (Figure 3.12b). Especially, the $LiFePO_4$||Li cell had a high discharge capacity after 90 cycles and showed stable voltage curves with slight polarization, indicating that the 4 M LiFSI/TMP electrolyte also formed a robust CEI to protect the cathode (Figure 3.12c,d). A

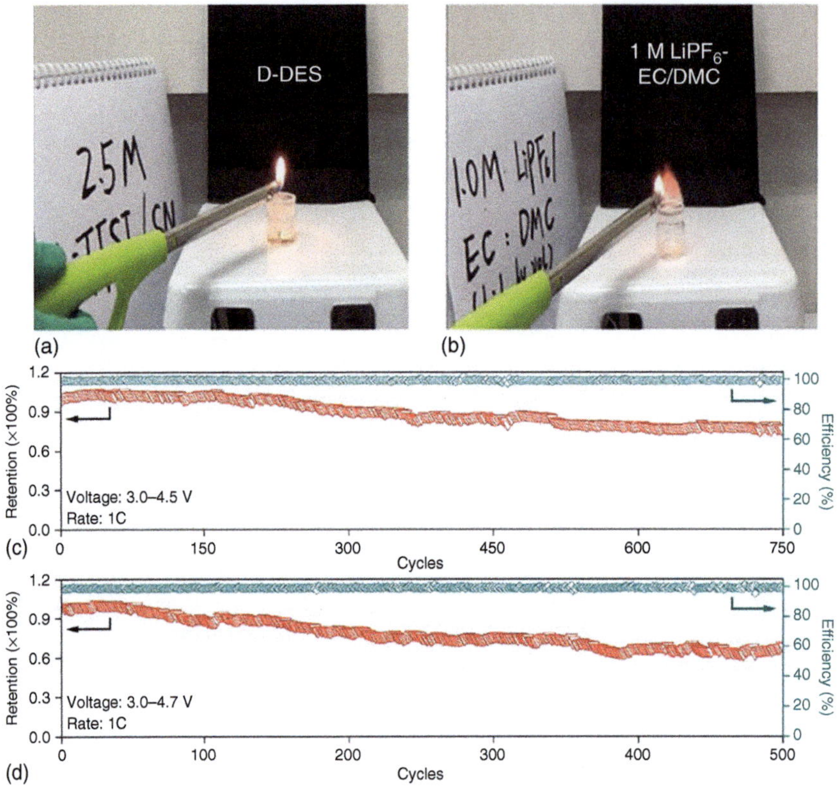

Figure 3.8 Flammability test of (a) D-DES and (b) 1 M $LiPF_6$-EC/DMC electrolytes. Discharge capacity retention of $LiCoO_2$||Li cells in D-DES electrolyte at a rate of 1 C with different voltage ranges: (c) 3.0–4.5 V and (d) 3.0–4.7 V. Source: Hu et al. [56] / with permission of American Chemical Society.

nonflammable high-concentration LiFSI-TEP electrolyte also demonstrated high chemical and electrochemical stability. Li metal plating/stripping showed high CE (>99.3%) in Li||Cu cells.

The unique character of HCEs coupled with fire-extinguishing properties contributes to develop safe and high-performance batteries. However, HCEs have their disadvantages such as high cost, high viscosity, and poor wettability [67, 68]. One possible solution is to introduce inert solvents to dilute HCEs. For example, Zhang and coworkers reported a nonflammable LHCE made by diluting the high-concentration LiFSI in TEPa with a diluent of bis(2,2,2-trifluoroethyl) ether (BTFE), which achieved dendrite-free Li deposition and enabled safe and high-performance Li||NMC622 batteries [69]. As shown in Figure 3.13, the cells with 1.2 M LiFSI in TEP/BTFE electrolyte exhibited ultrahigh stability compared to the baseline electrolyte. Xu et al. have reported the 1.2 M LiFSI in TEP/BTFE LHCE with 2.0% VC additives, which delivered a stable electrochemical performance in graphene||NCM batteries [70]. Through the transmission electron microscope (TEM), it was observed that a thin and robust SEI was formed on the anode

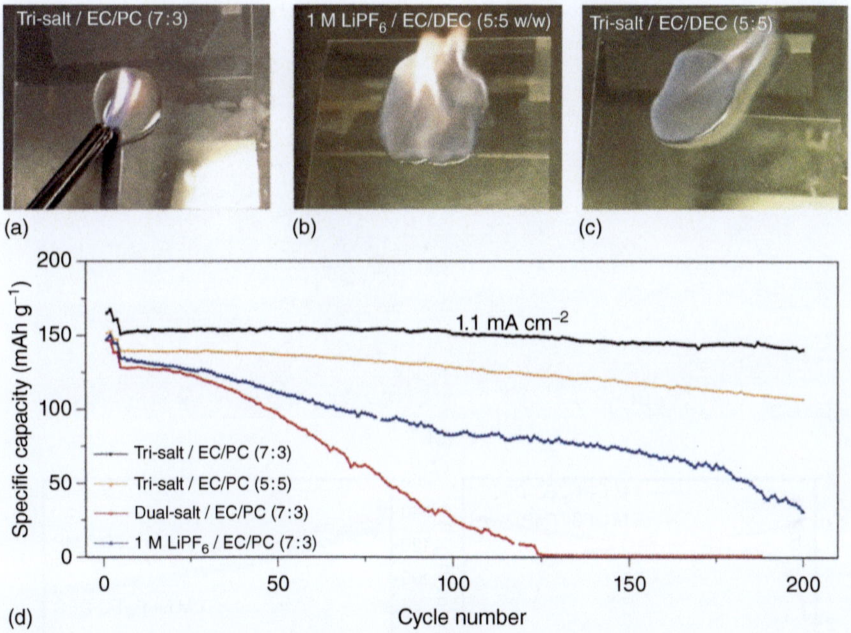

Figure 3.9 Flammability test of (a) tri-salt in EC/PC (7 : 3), (b) 1 M $LiPF_6$ in EC/DEC (5 : 5), and (c) tri-salt in EC/DEC (5 : 5). (d) Cycling performance with different electrolytes at 1.1 mA cm^{-2} after two cycles at 0.11 mA cm^{-2} and two cycles at 0.37 mA cm^{-2}. The capacity loadings of all cells are 1.1 mAh cm^{-2}. Source: Jin et al. [59] / with permission of American Chemical Society.

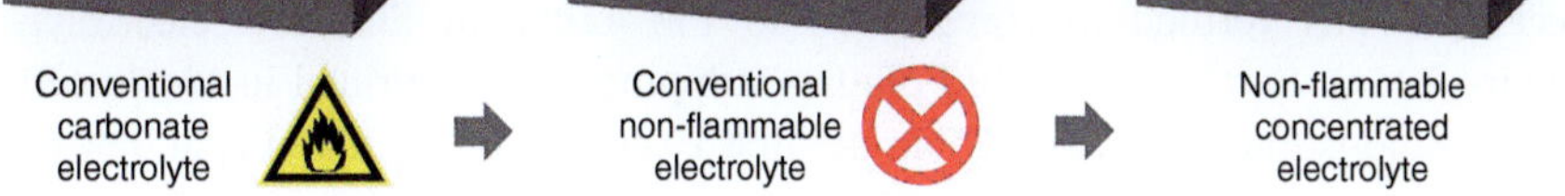

Figure 3.10 Intercalation behavior of cations (red spheres) into a carbonaceous anode in various electrolytes. Source: Reproduced with permission Wang et al. [64].

surface to enhance passivation ability, thus boosting the battery performance (Figure 3.14).

Ionic liquids (ILs) are high-safety solvents with intrinsic nonflammability [71, 72]. But LMBs with IL electrolytes have not yielded excellent performance due to unstable SEI formed from ILs on Li metal anodes [73, 74]. Considering the stable SEI formed in HCEs, the electrolytes consist of 1-ethyl-3-methylimidazolium

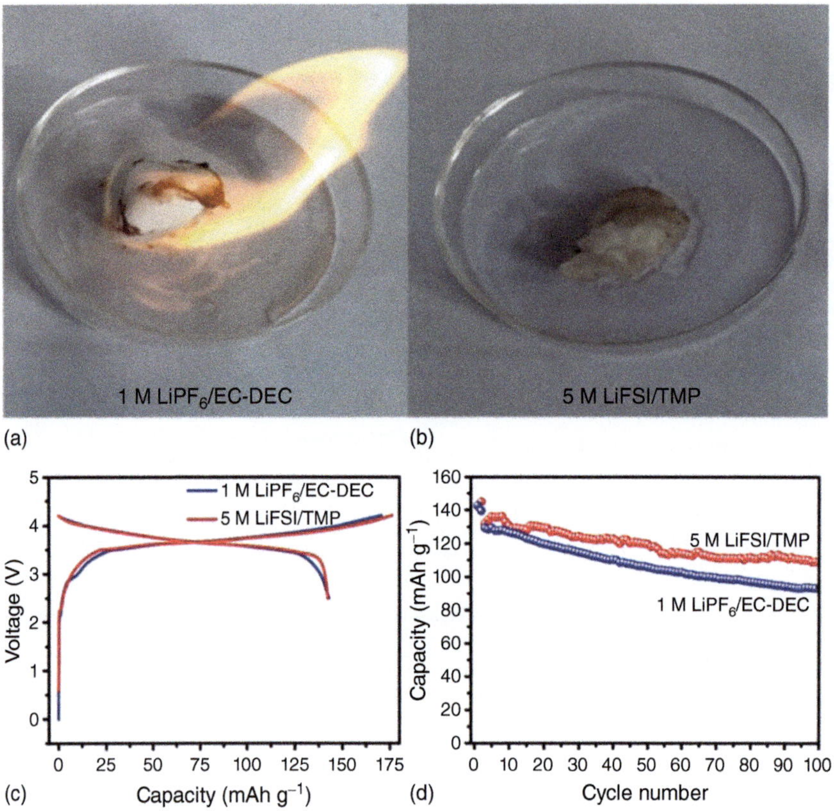

Figure 3.11 Flammability tests of (a) 1 M $LiPF_6$-EC/DEC and (b) the 5 M LiFSI/TMP electrolyte. (c) Initial charge–discharge curves and (d) cycling performance of the graphite||NMC cells with the 1 M $LiPF_6$-EC/DEC and 5 M LiFSI/TMP electrolytes. Source: Shi et al. [65] / with permission of Royal Society of Chemistry.

bis(fluorosulfonyl)imide ([EMIm]FSI), 5 M LiFSI and 0.16 M NaTFSI (donated as EM-5Li-Na) were reported by Dai et al. [75]. The nonflammability of the EM-5Li-Na electrolyte was verified in comparison to 1 M $LiPF_6$ in EC/DMC electrolyte (Figure 3.15a). Further, the CE Li plating/stripping was performed in Li–Cu cell, which demonstrated increased CE in the first 10 cycles and the stability at 99% for over 400 cycles (Figure 3.15b). The SEM images show that the deposited Li is bulk but not dendrites (Figure 3.15c). In high-concentration IL electrolytes, nonflammability and high electrochemical performance have been realized. Similarly, the LHCE of LiFSI as salt, *N*-methyl-*N*-propyl-piperidinium bis(fluorosulfonyl)imide ([PP13][FSI]) as solvent, and 1,1,2,2-tetrafluoroethyl-2,2,3,3-tetrafluoropropylether (HFE) as diluent have been designed [76]. Thermally stable ionic liquids enable the electrolyte flame-retardant ability and the high-concentration salt allows high Li plating/stripping performance (Figure 3.15d,e). All in all, similar electrolytes can be applied to LMBs and other battery systems to enhance their stability and safety.

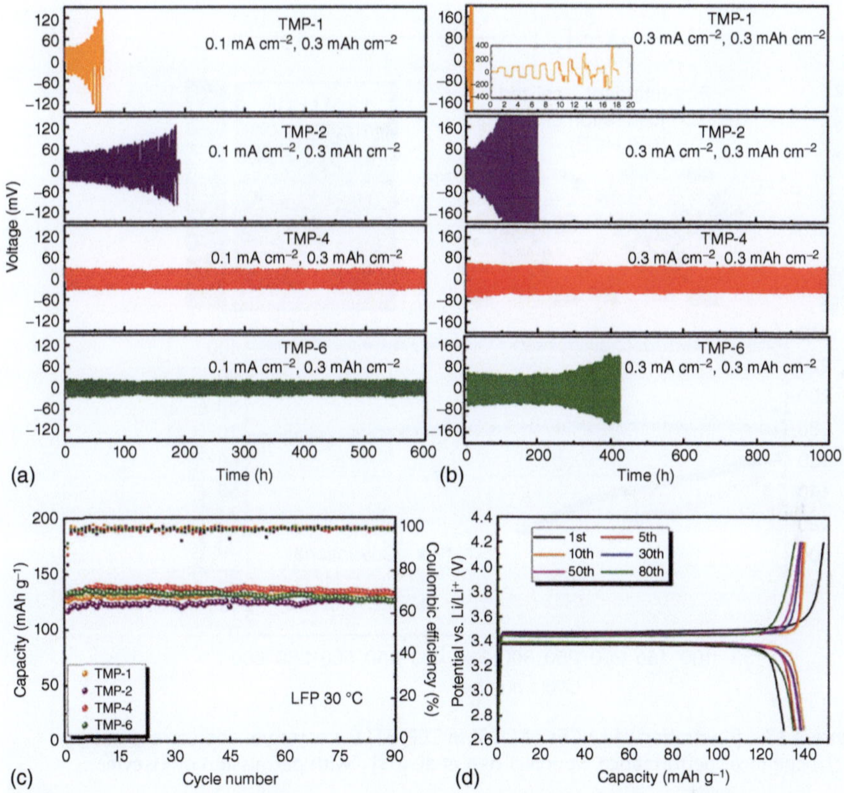

Figure 3.12 Li plating/stripping cycling test of Li symmetrical cells at the current densities of (a) 0.1 and (b) 0.3 mA cm^{-2}. (c) Cycling performance of the $LiFePO_4$||Li cells with different LiFSI concentrations at 0.1 C. (d) Charge–discharge curves of the $LiFePO_4$||Li cell with 4 M LiFSI/TMP electrolyte. Source: Reproduced with permission Zhang et al. [66].

3.5 Prospects

Development of high-safety and flame-retardant electrolytes is a key part in large-scale application of LMBs and other energy systems in the future. At present, using flame-retardant electrolytes can improve the safety performance of Li-ion, Li–O, and Li–S batteries, but at the cost of electrochemical performance. The development direction of nonflammable electrolytes in the future can be summarized as follows: (i) To develop multifunctional nonflammable additives or solvents of excellent interfacial compatibility with electrodes, which not only have nonflammable effect but also have interfacial film-forming function. (ii) To design new low-viscosity nonflammable ionic liquid-based electrolytes and to develop excellent non-flammable gel polymer electrolytes. (iii) To study the interface properties between nonflammable electrolytes and electrodes by in situ characterization techniques. (iv) To explore the redox decomposition mechanism to guide the optimization of nonflammable electrolytes. (v) To evaluate the flame-retardant effect of non-flammable electrolytes as well as the practicability of nonflammable electrolytes

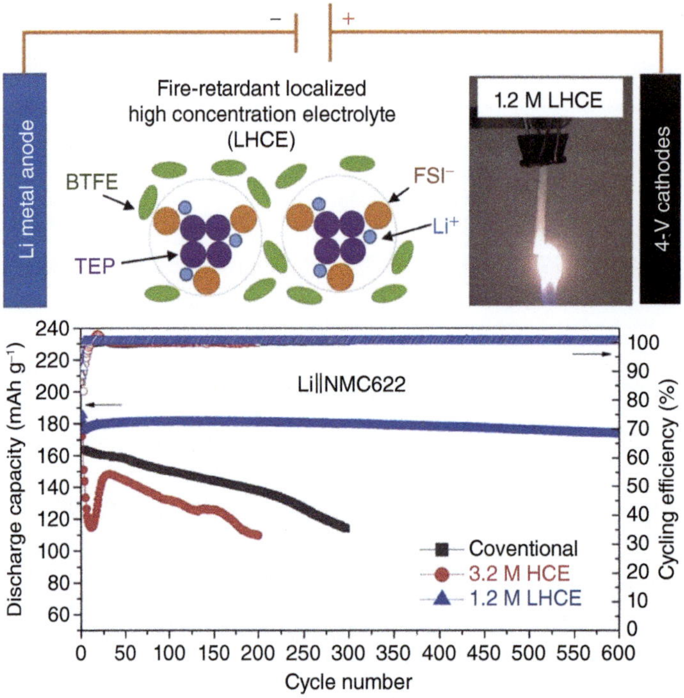

Figure 3.13 Illustration of LHCEs of LiFSI in TEP/BTFE electrolyte and corresponding electrochemical performance. Source: Chen et al. [69] / with permission of Elsevier.

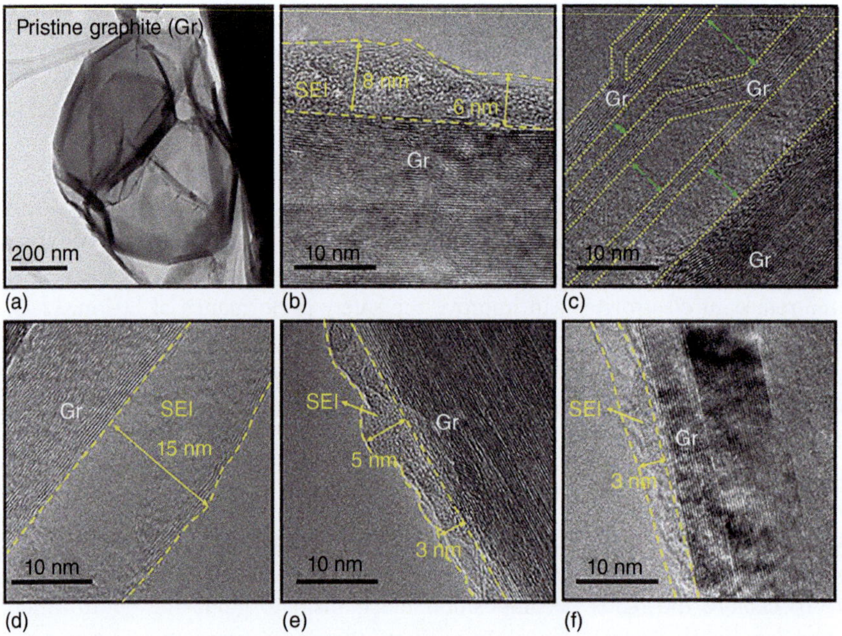

Figure 3.14 TEM images of (a) pristine Gr, (b) 1 M $LiPF_6$/EC-EMC, (c) 1 M $LiPF_6$/EC-EMC + 10% TEPa, (d) 1 M $LiPF_6$/EC-EMC + 10% TEPa + 2% VC, (e) 1.2 M LiFSI/TEPa-BTFE, and (f) 1.2 M LiFSI/TEPa-EC-BTFE. Source: Cao et al. [70] / with permission of American Chemical Society.

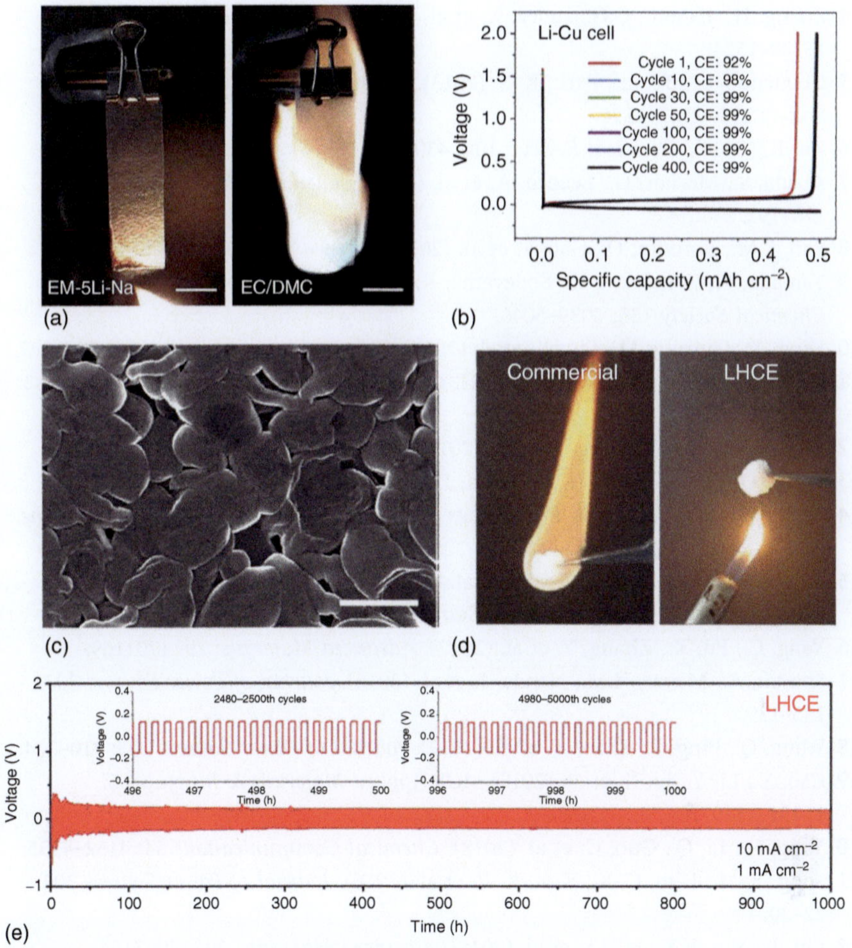

Figure 3.15 Flammability tests of (a) EM-5Li-Na electrolyte and 1 M $LiPF_6$ in EC/DMC (1 : 1 by vol). (b) Li plating/stripping profiles of Li–Cu cell with EM-5Li-Na electrolyte. (c) SEM image of plated Li on Cu foil. (d) Flammability tests 1.0 M $LiPF_6$ in EC:DMC:EMC (1 : 1 : 1 by vol) and IL/HFE LHCE. (e) Cycling performance of Li symmetric cell with IL/HFE LHCE at a current density of 10 mA cm^{-2} and a capacity of 1 mAh cm^{-2}. Source: Sun et al. [75] / with permission of John Wiley and Sons.

in large-capacity batteries such as electrochemical performance test, abuse test, thermal safety test.

References

1 Xue, P., Sun, C., Li, H. et al. (2019). *Advanced Science* 6: 1900943.
2 Armand, M. and Tarascon, J.M. (2008). *Nature* 451: 652–657.
3 Kim, H., Jeong, G., Kim, Y.-U. et al. (2013). *Chemical Society Reviews* 42: 9011–9034.

4 Zhang, H., Eshetu, G.G., Judez, X. et al. (2018). *Angewandte Chemie* 130: 15220–15246.

5 Goodenough, J.B. and Park, K.-S. (2013). *Journal of the American Chemical Society* 135: 1167–1176.

6 Xu, K. (2004). *Chemical Reviews* 104: 4303–4418.

7 Abada, S., Marlair, G., Lecocq, A. et al. (2016). *Journal of Power Sources* 306: 178–192.

8 Suo, L.M., Borodin, O., Gao, T. et al. (2015). *Science* 350: 938–943.

9 Yamada, Y., Furukawa, K., Sodeyama, K. et al. (2014). *Journal of the American Chemical Society* 136: 5039–5046.

10 Feng, X., Ouyang, M., Liu, X. et al. (2018). *Energy Storage Materials* 10: 246–267.

11 Todorov, Y.M., Aoki, M., Mimura, H. et al. (2016). *Journal of Power Sources* 332: 322–329.

12 Zhu, X., Wang, Z., Wang, Y. et al. (2019). *Energy* 169: 868–880.

13 Wang, Q., Jiang, L., Yu, Y., and Sun, J. (2019). *Nano Energy* 55: 93–114.

14 Hu, L. and Xu, K. (2014). *Proceedings of the National Academy of Sciences of the United States of America* 111: 3205–3206.

15 Wong, D.H., Thelen, J.L., Fu, Y. et al. (2014). *Proceedings of the National Academy of Sciences of the United States of America* 111: 3327–3331.

16 Yang, C., Fu, K., Zhang, Y. et al. (2017). *Advanced Materials*: 29, 1701169.

17 Sharafi, A., Meyer, H.M., Nanda, J. et al. (2016). *Journal of Power Sources* 302: 135–139.

18 Wang, Q., Ping, P., Zhao, X. et al. (2012). *Journal of Power Sources* 208: 210–224.

19 Cao, X., Li, Y., Li, X. et al. (2013). *ACS Applied Materials & Interfaces* 5: 11494–11497.

20 Yang, H., Li, Q., Guo, C. et al. (2018). *Chemical Communications* 54: 4132–4135.

21 Song, Y.-M., Kim, C.-K., Kim, K.-E. et al. (2016). *Journal of Power Sources* 302: 22–30.

22 Liu, L., Yin, Y.X., Li, J.Y. et al. (2018). *Advanced Materials*: 30, 1706216.

23 Xia, S., Wu, X., Zhang, Z. et al. (2019). *Chem* 5: 753–785.

24 Lin, D., Liu, Y., and Cui, Y. (2017). *Nature Nanotechnology* 12: 194–206.

25 Chen, J., Fan, X., Li, Q. et al. (2020). *Nature Energy* 5: 386–397.

26 Weber, R., Genovese, M., Louli, A. et al. (2019). *Nature Energy* 4: 683–689.

27 Wang, H., Matsui, M., Kuwata, H. et al. (2017). *Nature Communications* 8: 15106.

28 Fang, C., Li, J., Zhang, M. et al. (2019). *Nature* 572: 511–515.

29 Zhou, Y., Su, M., Yu, X. et al. (2020). *Nature Nanotechnology* 15: 224–230.

30 Cheng, X.B., Zhang, R., Zhao, C.Z., and Zhang, Q. (2017). *Chemical Reviews* 117: 10403–10473.

31 Xu, W., Wang, J., Ding, F. et al. (2014). *Energy & Environmental Science* 7: 513–537.

32 Belov, D.G. and Shieh, D.T. (2014). *Journal of Power Sources* 247: 865–875.

33 Chen, S., Wang, Z., Zhao, H. et al. (2009). *Journal of Power Sources* 187: 229–232.

34 Deng, K., Zeng, Q., Wang, D. et al. (2020). *Energy Storage Materials* 32: 425–447.

35 Hess, S., Wohlfahrt-Mehrens, M., and Wachtler, M. (2015). *Journal of the Electrochemical Society* 162: A3084–A3097.

36 Yao, X.L., Xie, S., Chen, C.H. et al. (2005). *Journal of Power Sources* 144: 170–175.
37 Zhang, L., Huang, Y., Fan, H., and Wang, H. (2019). *ACS Applied Energy Materials* 2: 1363–1370.
38 Wang, X., Yasukawa, E., and Kasuya, S. (2001). *Journal of the Electrochemical Society* 148: A1058.
39 Hyung, Y.E., Vissers, D.R., and Amine, K. (2003). *Journal of Power Sources* 119–121: 383–387.
40 Xiang, H.F., Xu, H.Y., Wang, Z.Z., and Chen, C.H. (2007). *Journal of Power Sources* 173: 562–564.
41 Azov, V.A., Egorova, K.S., Seitkalieva, M.M. et al. (2018). *Chemical Society Reviews* 47: 1250–1284.
42 Zeng, Z., Murugesan, V., Han, K.S. et al. (2018). *Nature Energy* 3: 674–681.
43 Yu, Z., Zhang, J., Wang, C. et al. (2020). *Journal of Energy Chemistry* 51: 154–160.
44 Wang, Q., Sun, J., Yao, X., and Chen, C. (2005). *Electrochemical and Solid-State Letters* 8: A467.
45 Matsumoto, K., Inoue, K., and Utsugi, K. (2015). *Journal of Power Sources* 273: 954–958.
46 Jiang, L., Liang, C., Li, H. et al. (2020). *ACS Applied Energy Materials* 3: 1719–1729.
47 Wu, L., Song, Z., Liu, L. et al. (2009). *Journal of Power Sources* 188: 570–573.
48 Nakagawa, H., Shibata, Y., Fujino, Y. et al. (2010). *Electrochemistry* 78: 406–408.
49 Todorov, Y.M., Fujii, K., Yoshimoto, N. et al. (2017). *Physical Chemistry Chemical Physics* 19: 31085–31093.
50 Shiga, T., Kato, Y., Kondo, H., and Okuda, C.-a. (2017). *Journal of Materials Chemistry A* 5: 5156–5162.
51 Yang, T., Li, S., Wang, W. et al. (2021). *Journal of Power Sources* 505: 230055.
52 Li, T., Zhang, X.-Q., Shi, P., and Zhang, Q. (2019). *Joule* 3: 2647–2661.
53 Arai, J. (2002). *Journal of Applied Electrochemistry* 32: 1071–1079.
54 Chung, G.J., Han, J., and Song, S.-W. (2020). *ACS Applied Materials & Interfaces* 12: 42868–42879.
55 Kim, Y.-S., Kim, T.-H., Lee, H., and Song, H.-K. (2011). *Energy & Environmental Science* 4: 4038–4045.
56 Hu, Z., Xian, F., Guo, Z. et al. (2020). *Chemistry of Materials* 32: 3405–3413.
57 Cha, J., Han, J.-G., Hwang, J. et al. (2017). *Journal of Power Sources* 357: 97–106.
58 Huang, S., Wang, S., Hu, G. et al. (2018). *Applied Surface Science* 441: 265–271.
59 Jin, T., Wang, Y., Hui, Z. et al. (2019). *ACS Applied Materials & Interfaces* 11: 17333–17340.
60 Yamada, Y., Yaegashi, M., Abe, T., and Yamada, A. (2013). *Chemical Communications* 49: 11194–11196.
61 Fan, X., Chen, L., Ji, X. et al. (2018). *Chem* 4: 174–185.
62 Liu, W., Li, J., Li, W. et al. (2020). *Nature Communications* 11: 3629.
63 Wang, J., Yamada, Y., Sodeyama, K. et al. (2016). *Nature Communications* 7 https://doi.org/10.1038/ncomms12032.

64 Wang, J., Yamada, Y., Sodeyama, K. et al. (2018). *Nature Energy* 3: 22–29.
65 Shi, P., Zheng, H., Liang, X. et al. (2018). *Chemical Communications* 54: 4453–4456.
66 Zhang, T., Li, Y., Chen, N. et al. (2021). *ACS Applied Materials & Interfaces* 13: 681–687.
67 Yamada, Y., Usui, K., Chiang, C.H. et al. (2014). *ACS Applied Materials & Interfaces* 6: 10892–10899.
68 Ding, Y., Yun, J., Liu, H. et al. (2014). *Pure and Applied Chemistry* 86: 585–591.
69 Chen, S., Zheng, J., Yu, L. et al. (2018). *Joule* 2: 1548–1558.
70 Cao, X., Xu, Y., Zhang, L. et al. (2019). *ACS Energy Letters* 4: 2529–2534.
71 Yang, Q., Zhang, Z., Sun, X.-G. et al. (2018). *Chemical Society Reviews* 47: 2020–2064.
72 Wang, Z., Sun, Y., Mao, Y. et al. (2020). *Energy Storage Materials* 30: 228–237.
73 Stark, J.K., Ding, Y., and Kohl, P.A. (2011). *Journal of the Electrochemical Society* 158: A1100.
74 Zhang, H., Qu, W., Chen, N. et al. (2018). *Electrochimica Acta* 285: 78–85.
75 Sun, H., Zhu, G., Zhu, Y. et al. (2020). *Advanced Materials* 32: 2001741.
76 Wang, Z., Zhang, F., Sun, Y. et al. (2021). *Advanced Energy Materials* 11: 2003752.

4

High-Voltage Electrolytes

4.1 Introduction

Following their successful commercialization in the early 1990s, rechargeable lithium ion batteries (LIBs) occupied the consumer market rapidly with the development of electronic cigarettes and portable electronics. Based on this commercialization process over the past three decades, LIBs have been regarded as a well-established and efficient electrochemical energy storage device in terms of power/energy density, life span, and wide working temperature ranges [1–3]. Currently, because of global environmental concerns and prospects for reducing the dependence on fossil fuels, a new large market for LIBs has been emerging. Electric vehicles (EVs), powered by battery, are expected to decrease air pollution of automobile exhaust gases, such as carbon monoxide, carbon dioxide, sulfur oxides, and particulate matters [4]. However, with the limited driving ranges, EVs are facing a tough competition with conventional automobiles. Among the various parameters, energy density is the most essential one, which determines the driving range for EVs. At present, the most common commercial LIBs are assembled by an intercalation/deintercalation type cathode and a graphite anode, offering gravimetric and volumetric energy densities of ~250–300 $W\,h\,kg^{-1}$ and ~600–700 $W\,h\,l^{-1}$, respectively. They can approach the 300 driving mile range, which cannot meet the target of 420 driving mile range. To meet this target, LIBs should possess a specific energy of 350 $W\,h\,kg^{-1}$ and an energy density of 750 $W\,h\,l^{-1}$ [5, 6].

Theoretically, energy density of a battery depends on its capacity and operating voltage (energy density = capacity × voltage). To improve the energy density, using cathode materials with high working voltage or electrode materials with high specific gravimetric capacity are achievable [7]. Looking back at the history of lithium-based batteries, the first cathode material was TiS_2, which was reported by Whittingham et al. 50 years ago. Not long after that, a series of chalcogenides were developed as cathode materials to store Li^+, with average voltages of less than 2.5 V (vs. Li/Li^+) [8, 9]. A major breakthrough was achieved in the 1980s. Goodenough and coworkers discovered a layered structured oxide cathode material, $LiCoO_2$, which can store Li^+ at about 4 V [10]. This discovery promoted the development

Liquid Electrolyte Chemistry for Lithium Metal Batteries: Design, Mechanisms, Strategies,
First Edition. Jianmin Ma.

of LIB technology. At the present time, the state-of-the-art cathode materials can be simply divided into three kinds. The first one is polyanion cathodes: $Li_xM_y(XO_4)_z$, where M = Fe, Mn, Co, V, etc. and X = P, Si, S, etc. $LiFePO_4$ is the most successful candidate of this kind of cathode material [11]. The second kind is Mn-based spinels, $LiMn_2O_4$. Low cost is the greatest advantage of this cathode material [12]. The third one is layered structured lithiated oxides, the successive generation of $LiCoO_2$ materials, such as nickel-based layered cathodes, $LiNi_xCo_yMn_zO_2$ and $LiNi_xCo_yAl_zO_2$ ($x + y + z = 1$) [13, 14]. Among these cathode materials, $LiCoO_2$, $LiFePO_4$, $LiMn_2O_4$, and nickel-based layered oxides have been widely commercialized. However, due to low operating voltage or limited capacity, $LiFePO_4$ and $LiMn_2O_4$ are not regarded as suitable cathode materials for high-energy batteries. In recent years, massive efforts have been devoted to design and optimization of novel cathode materials with high working voltage (>4 V vs. Li^+/Li) and/or large specific capacity (>200 mAh g^{-1}). The reported promising candidates of cathode materials include lithium-rich layered oxides ($Li_{1+x}M_{1-x}O_2$, M = Mn, Ni, Co, etc.), high-voltage spinel oxides ($LiNi_{0.5}Mn_{1.5}O_4$), high-voltage polyanionic compounds, high-voltage $LiCoO_2$, and Ni-rich layered oxides [15, 16]. Some developments have been achieved. For example, by doping metal cations to suppress the phase transition from H_1 to O_3, the modified $LiCoO_2$ can be charged to 4.5 V or even higher [17]. With successful commercialization of layered lithiated oxides with low Ni content (e.g. $LiNi_{1/3}Co_{1/3}Mn_{1/3}O_2$), it is believed that the Ni-rich compositions (Ni content >0.6) can be optimized to reach the target of 300 W h kg^{-1} in the near future. Besides, lithium-rich layered oxides can offer a high gravimetric specific capacity of about 250 mAh g^{-1} and manganese-based spinel cathodes have the ability of operating at 4.7 V. For the anode side, the most used commercialized electrode material is graphite, which was realized by Sony in the early 1990s [18]. Graphite can store Li^+ at a low voltage of ~0.1 V (vs. Li^+/Li). However, theoretical specific capacity of this carbonaceous material is 372 mAh g^{-1}, limiting the energy density of batteries with graphite anode. In the past decade, advanced anode materials with high capacity were reported, such as transition metal oxides (Fe_2O_3, SnO_2) [19] and silicon (4200 mAh g^{-1}) [20]. However, their relatively high operating voltages also limit the improvement of energy density of LIBs, and their structural stability is worrying. Metallic lithium has high theoretical specific capacity (3862 mAh g^{-1}, around 10 times greater than that of graphite) and lowest Li^+ storage potential (−3.04 V vs. the standard hydrogen electrode) [21]. As the initial negative electrode material, metallic lithium had a short application history in the 1980s. However, due to unacceptable safety risks caused by uneven Li electrochemical deposition, metallic lithium was dismissed. Benefited by the development of materials science, metallic lithium has been refocused in recent years. Novel current collector and electrolyte systems have been developed to overcome the problem of uneven Li electrochemical deposition in part. Researches about lithium metal batteries (LMBs) have been booming nowadays.

To develop power batteries with high energy density is systems engineering. Improvement in the working voltage will bring many unexpected impacts. First, when the working voltage is high, conventional carbonate electrolytes cannot withstand. Oxidative decomposition of the electrolyte will lead to an increase in the internal resistance and formation of gas (CO_2), causing safety hazards. Although some new electrolyte systems have been reported, such as sulfone or nitrile-based electrolytes, their chemical/physical properties and compatibility with cathode/anode/separator should be investigated systematically. Cathode materials may go through some different reaction processes, which affects their structure and electrochemical stability. Moreover, some inert portions may become non-inert at a high voltage. It has been reported that the conductive carbon black and aluminum current collector can react with electrolytes at a voltage of more than 4.8 V [22]. These issues are as important as the development of cathode electrode materials for high-voltage batteries. Considering rapid progress in research on those issues, it is time to review the electrolyte systems and high-voltage battery issues. This section discusses the state-of-the-art electrolyte systems and additives and the formation mechanism of cathode electrolyte/electrode interphase (CEI) and interphase stability of cathodes.

4.2 The General Implications of High-Voltage Electrochemical Operation

4.2.1 Electrochemical Stability and Voltage Window for Electrolytes

Benefited by the development of computational chemistry, theoretical calculation can be used to predict the voltage window of an electrolyte. As shown in Figure 4.1, electrochemical oxidation and reduction potentials of an electrolyte are determined by its highest occupied molecular orbital (HOMO) and lowest unoccupied molecular orbital (LUMO), respectively [23]. In the ideal case, the electrolyte only acts as a chemically inert medium to enable ions transportation between the cathode and anode. It requires the electrolyte should operate within the voltage window of HOMO and LUMO [24–26]. However, in fact, the Li^+ storage potential of the anode electrode is almost lower than the reduction potential of conventional carbonate electrolytes. The reduction of electrolyte has complex influences on the electrochemical performance of LIBs and LMBs. On the positive side, the formed solid/electrolyte interphase (SEI) is a stable passivation film, which can prevent the further reduction of electrolyte and cointercalation of Li^+ and organic solvent molecules for graphite anode [27]. On the negative side, the formation of SEI film will consume Li^+ and solvent, causing the low coulombic efficiency (CE) and increasing resistance. Some anode materials will undergo a huge volume change during Li^+ intercalation/deintercalation, such as Si and metal oxide anodes [28]. Stable SEI cannot form on surfaces of this kind of anode material. The problems of low CE and high resistance will aggravate. Therefore, the importance of SEI has spurred extensive research efforts to understand its formation and optimize its functioning mechanisms.

As for the cathode side, commercially established conventional electrode materials ($LiCoO_2$, $LiFePO_4$, and $LiMn_2O_4$) do not operate at high voltage, at ~4 V. It is

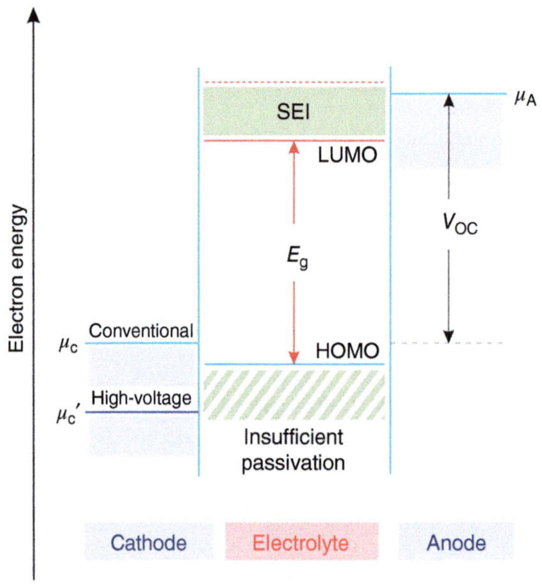

Figure 4.1 A schematic of the open-circuit energy diagram, showing the relationship between the voltage window and the formation condition of the electrolyte/electrode interphase at the two electrodes surface. Source: Reproduced with permission Li et al. [23].

generally acknowledged that the oxidation stability limit of nonaqueous carbonate electrolytes is ~4.3 V (vs. Li^+/Li) [29]. Therefore, due to lack of thermodynamic driving force for electrolyte oxidation, the formation and stability of the CEI is not perceived crucial for structural stability of cathode electrode and cycle life of batteries. Changes appear in recent years. As mentioned at the beginning of this section, novel high-voltage cathode materials, including spinel $LiNi_{0.5}Mn_{1.5}O_4$, layered Ni-rich oxides, polyanionic cathodes, and modified $LiCoO_2$, show the Li^+ intercalation/deintercalation potential beyond the electrolyte oxidation limits [30]. The electrolyte and electrode compatibility becomes very important for high-voltage batteries. Although some recent works attempted to use organic solvents with lower HOMO values to prepare novel electrolytes, such as sulfones, nitrile, and fluorocarbonates, the chemical/physical properties and compatibility with anode electrodes of those as-prepared electrolytes are suspicious and should be systematically studied [31]. Moreover, the reasons for capacity decline of high-voltage batteries are more complex, which can be categorized into parasitic electrolyte oxidation, mass loss and crystal structure reconstruction of cathode materials, breakdown of current collector, spontaneous reactions involving conductive carbon additives, and others [23]. In 4.2.2–4.3.4 sections, the general mechanisms of capacity fade for high-voltage batteries are discussed and the respective strategies for various cathode materials are reviewed.

4.2.2 Parasitic Electrolyte Oxidation and Formation of CEI

For the present LIBs/LMBs electrolyte systems, the dilute $LiPF_6$ dissolved in carbonate-based organic solvents (such as ethylene carbonate, EC; ethyl methyl carbonate, EMC; dimethyl carbonate, DMC; diethyl carbonate, DEC) is the skeleton formulation. As mentioned earlier, the oxidation potential limits of those nonaqueous carbonate-based electrolytes are generally located at 4.3 V (vs. Li/Li^+). When the potential becomes higher than this value, irreversible oxidation decomposition will take place, triggered by various factors, including elevated temperature, trace moisture, and catalytic metallic ions on the surface of the cathode electrode [32]. Therefore, the mechanism of the electrolyte breakdown and subsequent polymerization on the cathode electrode can be described in the following two modes: (i) At highly delithiated states of cathode materials, decomposition will be catalyzed by the nucleophilic attack of the active cathode material, as shown in Figure 4.2a,b [32, 35]. (ii) With trace moisture, the breakdown can be triggered by the attack of

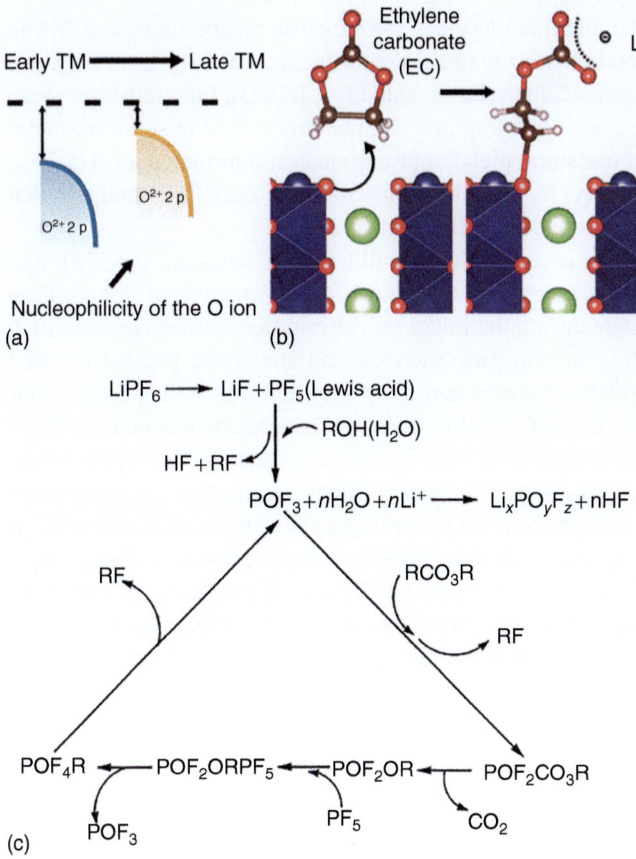

Figure 4.2 (a) A schematic showing the surface reactivity of a layered transition metal (TM) oxide cathode at a highly delithiated state, nucleophilicity of the oxygen atom on the surface layer with different transition metal ions. (b) The ring-opening oxidation process of an EC molecule on the surface of the layered TM oxide cathode. Source: Reproduced with permission Fan et al. [33]. (c) The mechanism for the auto-catalytic decomposition of $LiPF_6$-containing electrolyte with moisture. Source: Reproduced with permission Xu et al. [34].

acidic products of LiPF6 hydrolysis, as displayed in Figure 4.2c [34, 36]. This process produces the CEI film consisting of highly complex organic and inorganic species on the cathode surface, as well as gaseous products. To distinguish the composition of the CEI, several characterization techniques have been used, including ex situ X-ray photoelectron spectroscopy (XPS), Fourier transform infrared (FTIR) spectroscopy, in situ neutron reflectometry (NR), solid-state nuclear magnetic resonance (NMR), etc. [37–39]. For example, Chen and coworkers investigated the CEI on $LiNi_{1/3}Co_{1/3}Mn_{1/3}O_2$ cathode by time-of-flight secondary ion mass spectrometry (TOF-SIMS), and the results are shown in Figure 4.3a,b [40]. As can be seen, the signals of C–H–O organic compounds, P-compounds, transition metal fluorides/oxides, and LiF can be detected by TOF-SIMS. Li_2CO_3, organic species, and P-compounds are mainly distributed on the surface of the CEI, whereas the metal fluorides/oxides are mainly located in the inner layer. Besides, LiF species uniformly spread on both the outer and inner layers of the CEI. The organic/inorganic hybrid structure of the CEI is schematically illustrated in Figure 4.3c [41]. Generally, it is believed that the inorganic CEI species include Li_2O, Li_2CO_3, LiF, transition metal fluorides, $Li_xPO_yF_z$; the organic interphase components consist of semicarbonates ($ROCO_2M$, M = Li and transition metal), polycarbonates, fluorinated alkyl (RCF_x), alkoxides, etc. At first glance, the chemical composition of the CEI is similar to that of the SEI on graphite or metallic lithium anode [42].

With the development of research, the differences between the CEI and SEI are recognized in multiple aspects. First, the formation of the SEI on graphite/lithium/Si anode during the initial reductive process always leads to high irreversible capacity. By contrast, Kasnatscheew and coworkers pointed out that only the electrolyte oxidation process contributes limited irreversible capacity for the Li||$LiNi_{1/3}Co_{1/3}Mn_{1/3}O_2$ half-cell [43]. In detail, the capacity loss of 36 mAh g^{-1} is induced by kinetic limitations of the impeded lithiation reaction during discharge. Through a constant-potential step at the end of discharge, most of the capacity can be recovered. Second, for commercial LIBs with graphite anode, the formed SEI is passivated enough to prevent the further parasitic reductive reaction of electrolyte. However, due to incomplete formation, the passivation effect of the CEI is typically less than that of SEI on graphite. After the formation of CEI during the initial cycle, oxidative electrolyte breakdown will still take place throughout subsequent cycles. As a result, the parasitic oxidation reaction at high voltage brings the consumption of electrolyte and Li^+, increasing impedance, excessive gas production, transition metal ion diffusion, and even premature cell death.

The gaseous production of parasitic electrolyte decomposition is one of the main reasons of battery swelling. Both electrolyte oxidation and reduction decomposition can produce gaseous species. Figure 4.4a,b shows the shape change of a pouch cell before and after gas generation [43]. The swelling caused by gas generation not only can increase the interphase resistance of the battery but also can cause safety issues. The predominant gaseous product of electrolyte oxidation decomposition is CO_2. Kasnatscheew and coworkers used the on-line MS to measure the gas generation on $LiNi_{1/3}Co_{1/3}Mn_{1/3}O_2$ cathode [44]. As can be seen from Figure 4.4c, the CO_2 production occurs at above 4.6 V (vs. Li^+/Li). At high temperatures, the gas-generating

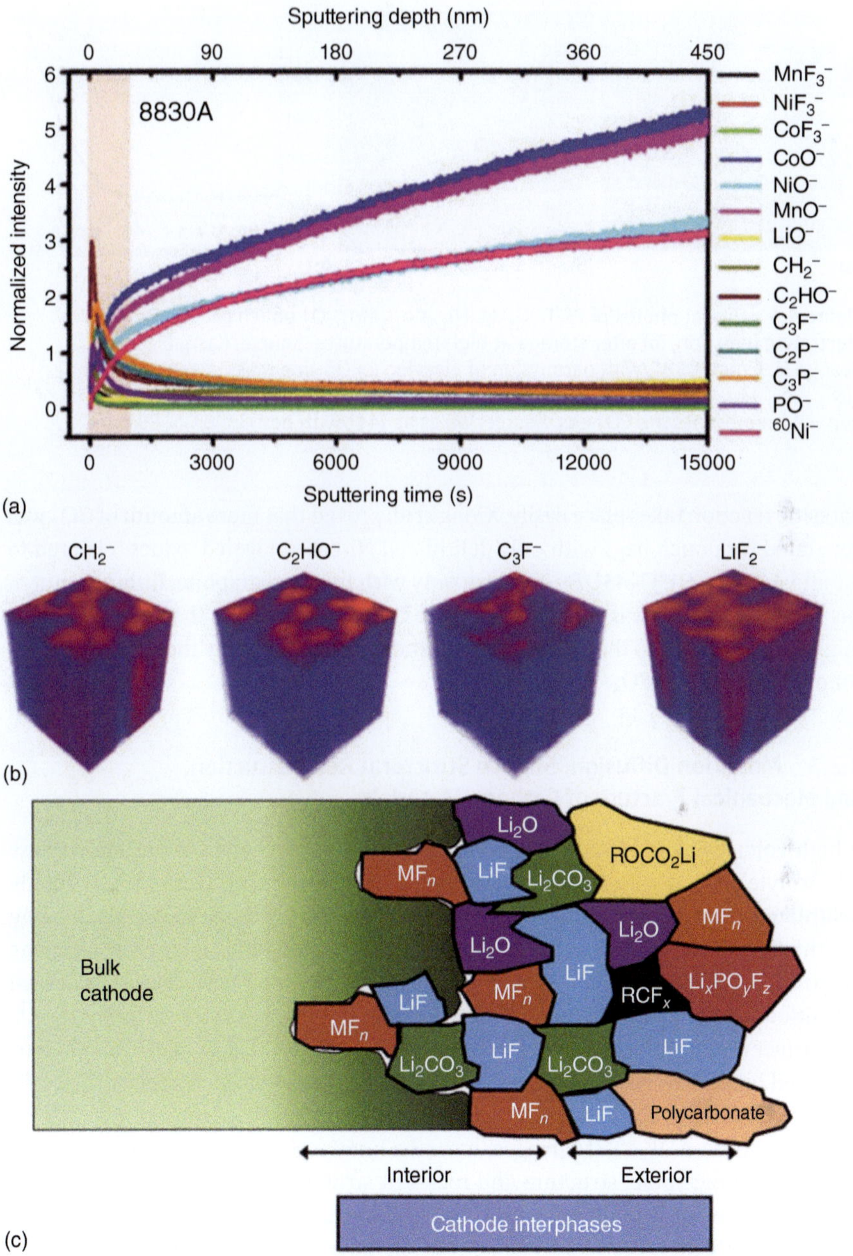

Figure 4.3 Time-of-flight secondary ion mass spectrometry (TOF-SIMS) characterizations of CEI on $LiNi_{1/3}Co_{1/3}Mn_{1/3}O_2$ cathode, in $LiPF_6$-EC/EMC/DMC electrolyte: (a) the normalized intensity of various ion fragments at different depths for the CEI, formed in the electrolyte with fluoroethylene carbonate (FEC) and $LiPO_2F_2$ additive. (b) 3D render of composition and concentration distribution of the CH_2^-, C_2HO^-, C_3F^-, and LiF_2^- ion fragments in the CEI formed without any additive. Source: Reproduced with permission Chen et al. [40].
(c) A schematic illustration of the organic/inorganic CEI on cathode at high voltage. Source: Li et al. [41]. Licensed under CC BY 4.0.

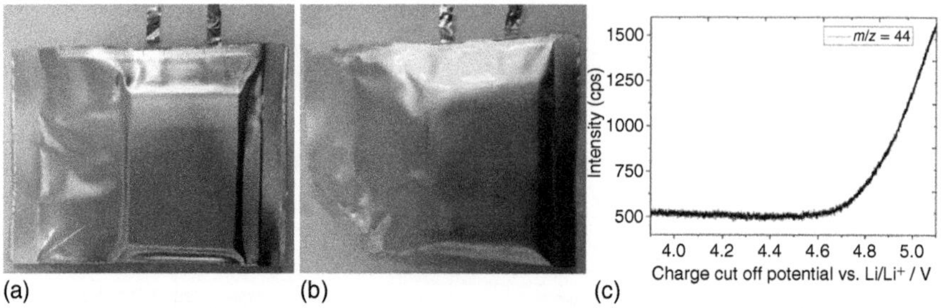

Figure 4.4 Digital photos of $Li_4Ti_5O_{12}||LiNi_{1/3}Co_{1/3}Mn_{1/3}O_2$ pouch cell. (a) Before electrolyte injection, (b) after storage at high temperatures. Source: Kasnatscheew et al. [43], Figure 01BC/with permission of Elsevier. (c) On-line mass spectroscopy for a $LiNi_{1/3}Co_{1/3}Mn_{1/3}O_2$ during a linear sweep with a scan rate of 0.034 mV s^{-1}. The signal of $m/z = 44$ represents the CO_2 gas. Source: Wu et al. [44]/with permission of Elsevier.

parasitic reaction takes place easily. Xiong et al. proved that more amount of CO_2 was generated for pouch bags with only delithiated Ni-based layered oxides (charged to 4.2 and 4.4 V) at 60 °C [45]. By contrast, only with lithiated graphite, little amount of gas was generated. It was concluded that the CO_2 produced from the cathode surface might have diffused to the graphite anode and was consumed by the reductive environment to form Li_2CO_3.

4.2.3 Metal Ion Diffusion, Surface Structural Reconstruction, and Mechanical Fracture of Cathode Materials

A high-voltage environment not only brings the parasitic reaction in the liquid phase (electrolyte) but also leads to the side reaction in solid state (cathode). A highly delithiated state at high voltage is an unstable state of cathode materials, leading to lithiation reaction and side reactions of transition metal ions migration, shearing of atomic layers in the lattice upon Li removal. At the interface region, the crystal transition is more complex than in the bulk phase [46, 47]. As shown in Figure 4.5, the surface structural rearrangement of $LiNi_{0.4}Mn_{0.4}Co_{0.18}Ti_{0.02}O_2$ particle was characterized by atomic-resolution ADF-STEM [48]. Figure 4.5a shows the surface of a $LiNi_{0.4}Mn_{0.4}Co_{0.18}Ti_{0.02}O_2$ particle after one complete cycle between 2.0 and 4.7 V. By analyzing those ADF-STEM images, it can be known that the rearrangement layer consists of some spinel structure and rock-salt structure crystals. The thickness of the rearrangement layer is influenced by two factors. At first, the charge–discharge process can thicken the rearrangement layer. Then, different crystal orientations also have impact on the thickness. Generally, along the lithium diffusion channels (R-3m [50] zone axis), the formed rearrangement layer is thicker than along other orientations. It suggests that the reconstruction process is severe in the region to transport Li^+. In some extreme cases, a dangling layer full of cubic rock-salt crystals can be observed, as exhibited in Figure 4.5b. This loosely attached layer is susceptible to ion dissolution in electrolyte, especially Mn^{2+} dissolution. Similarly, the crystal structure rearrangement takes place in the Li-rich cathode materials. Cho's group

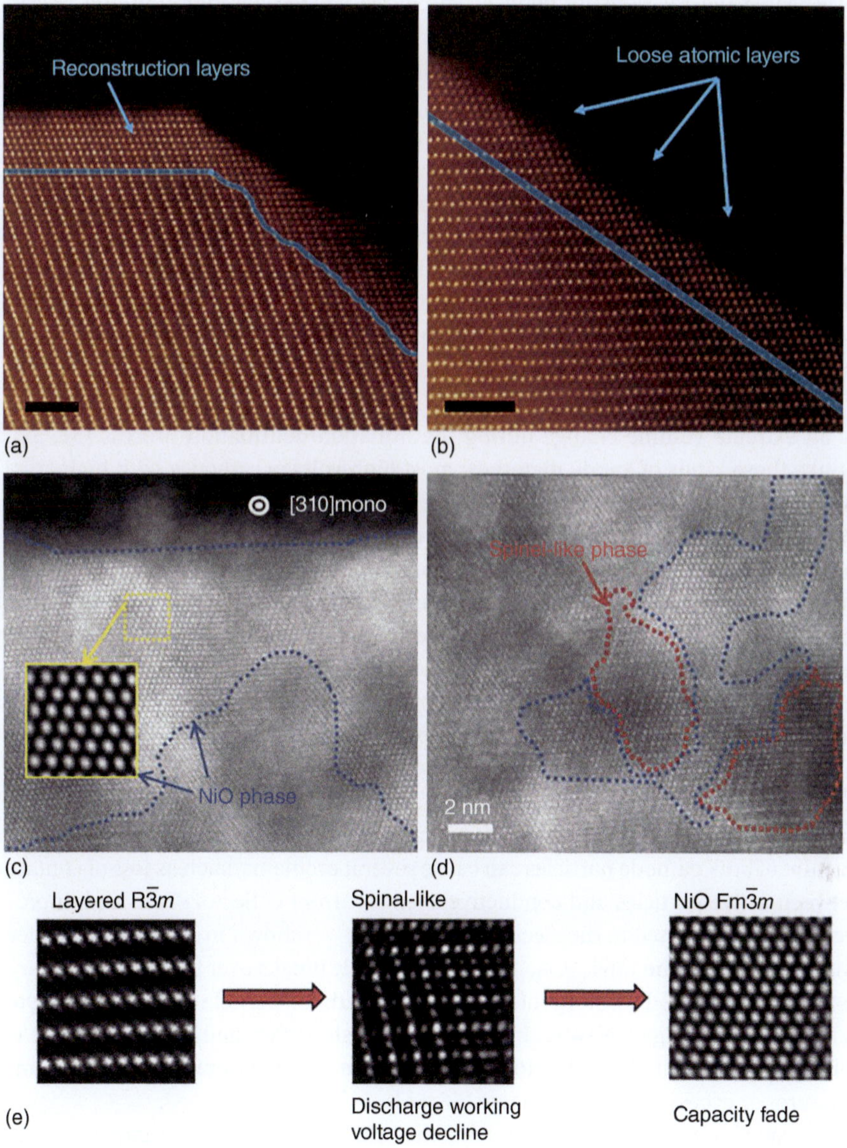

Figure 4.5 Atomic-resolution ADF-STEM images of $LiNi_{0.4}Mn_{0.4}Co_{0.18}Ti_{0.02}O_2$ cathode particles. (a) After exposure to LiPF6-EC/DEC electrolyte for 30 h. (b) The loose atomic $LiNi_{0.4}Mn_{0.4}Co_{0.18}Ti_{0.02}O_2$ layer on particles, after the initial cycle between 2 and 4.7 V. The scale bars are 2 nm in the first 2 images. Source: Yan et al. [48]/with permission of Springer Nature. (c, d) STEM-HAADF images of $0.4Li_2MnO_3$-$0.6LiNi_{1/3}Co_{1/3}Mn_{1/3}O_2$ particles, along $[310]_{mono}$ (=$[110]_{spinel}$) after 100 cycles. The regions marked by the blue and red lines indicate the disordered transition metal ion rearrangement phase and spinel-like phase. The two images have the same scale bar. (e) Crystal phase transition process of $0.4Li_2MnO_3$-$0.6LiNi_{1/3}Co_{1/3}Mn_{1/3}O_2$ cathode material during cycles at high voltage: from the R-3m layered to the Fm-3m cubic rock-salt structure. Source: Oh et al. [49]/with permission of Wiley.

investigated the relationship between discharge voltage deterioration and crystal structure transition [49]. As shown in Figure 4.5c,d, after 100 cycles between 2.0 and 4.8 V, the NiO phase (*Fm*3*m* cubic rock-salt structure) and spinel-like phase can be observed in the $0.4Li_2MnO_3{-}0.6LiNi_{1/3}Co_{1/3}Mn_{1/3}O_2$ cathode. The crystal structure transition is caused by the metal ion rearrangement. In detail, when charged to 4.6 V or higher voltage, many Li^+ vacancies are generated at the surface of Li-rich cathode particles. Then, transition metal ions migrate to the vacancies and the transition metal ion vacancies are also formed. The continuous crystal phase transition from the surface to the bulk phase is triggered by two kinds of vacancies. As shown in Figure 4.5e, the spinel-like phase is the intermediate phase between the layered and rock-salt phase. The two impurity phases, spinel-like and cubic rock-salt, cause deterioration in the discharging working voltage and capacity decay.

For some anode materials, such as Si, Fe_2O_3, SnO_2, mechanical fracture is caused by an extreme volume change during the lithiation/delithiation process [51, 52]. Unlike those kinds of anode materials, most high-voltage cathodes with high crystallinity only undergo a marginal volume change, even at the highly delithiated state. However, the mechanical fracture still occurs on high-voltage cathodes. The most common reason for structural crack on voltage cathodes is anisotropic microstrains along different crystal axis directions. For example, $LiCoO_2$ will suffer from the phase transitions from H3 to H1-3 phase and H1-3 to O1 phase at 4.55 and 4.63 V, respectively. A crystal structure shrinkage accompanies these phase transitions. During the first transition (H3 to H1-3 phase), inhomogeneous strain also builds up with the c-axis contracted by 6%, and the a-axis expanded by 0.6%. This strain difference is one of the main reasons for serious structure collapses of $LiCoO_2$ at high voltage [53]. Apart from phase transition, many other factors can be attributed to microstrains, such as ion vacancies, local composition variation, and cation mixing. Mechanical fracture of bulk cathode particles can cause several problems, such as loss of contact between active particles and conductive agents/current collector, fresh and unprotected surface exposed to the electrolyte, and so on. As shown in Figure 4.6a,b, after cycled up to 4.2 V, the $LiNi_{0.6}Co_{0.15}Mn_{0.25}O_2$ particle breaks over the grain boundary [54]. The formed CEI cannot protect the broken cathode particle and the electrolyte penetrates into the gap. New CEI forms on the fresh surface and more electrolyte is consumed, leading to low CE. Mechanical fracture will deteriorate at elevated temperatures or excessive deep discharge [55, 56].

The parasitic reaction also generates some soluble products, for example, soluble transition metal ions. The earliest notable soluble transition metal ion is probably Mn^{2+} cation from manganese-based cathode materials [57–59]. Mn^{2+} is generated from the disproportionation reaction of Mn^{3+}. The active material loss due to dissolution of transition metal ions commonly occurs for most positive electrode materials, in the $LiPF_6$/carbonate electrolyte. This phenomenon is more pronounced at high voltage. Many researchers have devoted their efforts to understand the adverse effects of soluble transition metal ions. For example, a research work pointed out that for the "5-V" $LiN_{0.5}Mn_{1.5}O_4$, only a loss of 1–2 at% Ni/Mn species can cause a huge capacity fade of 50% of the entire cell after only 100 cycles at ambient temperature [60]. Interestingly, dissolution of Mn^{2+} and Ni^{2+} ions not only

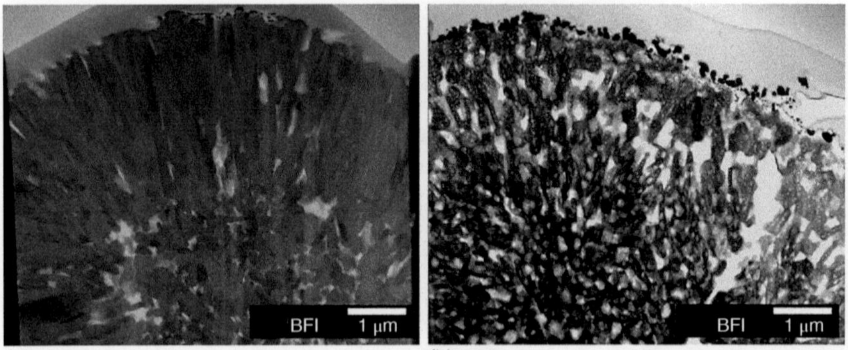

(a) (b)

Figure 4.6 Cross-sectional TEM image of Ni-rich oxide cathode particles, (a) at the pristine state and (b) after charged to 4.4 V at 25 °C. Source: Lee et al. [54]/with permission of American Chemical Society.

decreases the stability of $LiN_{0.5}Mn_{1.5}O_4$ cathode but also injures the passivation of SEI on graphite anode [57]. This is probably because the metal species will be reduced at the surface of anode electrode automatically. The deposited transition metal particles in the interior SEI catalyze continuous parasitic electrolyte reduction, leading to the sustained SEI, which impedes Li^+ diffusion. Preventing the dissolution of transition metal ions is a key process in the optimization of high-voltage batteries.

4.2.4 Instability of Other Cell Components at High Voltage

In general, some components of batteries are regarded as inert, such as conductive carbon additives, separator, current collector, and containers. However, according to some recent works, the inert nature of those components is suspicious at extreme high voltages. First, aluminum (Al) foil is the most common current collector for positive electrodes. With $LiPF_6$ or $LiBF_4$, Al could be passivated due to the formation of AlF_3. At elevated temperatures or high-voltage environments, side reactions will still take place, such as sustained dissolution of the Al^{3+} cations with electrolyte components. Li and coworkers investigated the stability of the Al current collector at high voltage [61]. As shown in Figure 4.7a, at 4.9 V, although the edge of the Al current collector is blurred, there exists no crack. When the voltage is increased to 5.3 V (Figure 4.7b), mechanical failure emerges. The electrochemical test shows the same result. As can be seen from Figure 4.7c, below 5 V, there is no parasitic reaction. At 5.3 V, a large residual capacity emerges, meaning the reactivity change of the Al current collector at this high voltage. In addition, many works reported that in dilute LiTFSI-based electrolytes, Al is not passivated as in $LiPF_6$ and $LiBF_5$-based electrolytes. The corrosion of Al often takes place at a relatively low voltage. According to the study by Matsumoto's group, the Al current collector will be oxidized immediately from 3.7 V in the 1 M LiTFSI/ EC-DEC (3 : 7, by volume) electrolyte [62]. As shown in Figure 4.7d, after the oxidation corrosion, many pores emerge on the Al surface. Increasing the concentration of LiTFSI can protect the Al current

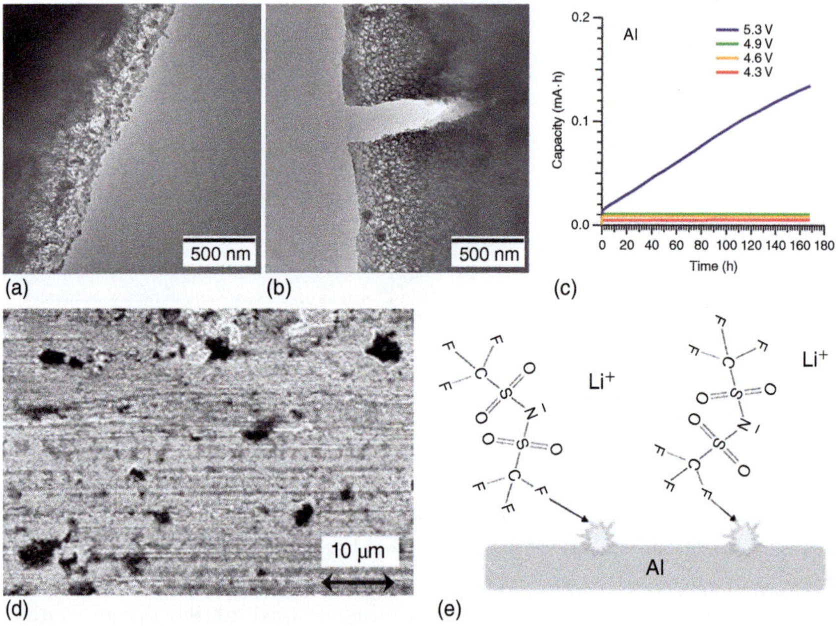

Figure 4.7 TEM images of the Al stored at (a) 4.9 V and (b) 5.3 V. (c) Residual current (top) and capacity (bottom) of lithium/Al cells stored at different potentials. Source: Li et al. [61]/with permission of IOP Publishing. (d) SEM image of Al foil after charged in diluted electrolyte. (e) Schematic diagram showing the corrosion of LiTFSI to the Al current collector. Source: Matsumoto et al. [62]/with permission of Elsevier.

collector even at 5 V, probably due to the formation of LiF-rich passivation layer, as illustrated in Figure 4.7e.

As shown in Figure 4.8a, the parasitic reaction of conductive carbon additive takes place at a lower voltage than that of Al current collector in the $LiPF_6$/carbonate electrolyte [61]. The relatively high reactivity of conductive carbon additive can be simply attributed to two main factors. Firstly, as conductive additives, carbon black, carbon nanotubes, and graphene all have an extremely larger surface area than that of cathode materials. Although the mass ratio of conductive carbon additives in the positive electrode is low, less than 10% in most cases, they contribute most of the surface area. Due to the manufacturing or storage process, some oxygen-containing functional groups exist on the surface of carbonaceous materials, such as carbonyl, carboxyl, lactone, and hydroxyl groups, as displayed in Figure 4.8b [63]. At high voltage, those groups tend to catalyze decomposition of the electrolyte. According to previous studies, surface deposition of hydrocarbons, carbonates, Li_2CO_3, LiF, etc., can be detected on the surface of carbon cathode, with partial decomposition on charging to 4.3–4.5 V. To increase graphitization and decrease the content of chemical functional groups on conductive carbon additives are beneficial to high-voltage batteries. Secondly, other electrochemical reactivity of conductive carbon additives is the anion insertion reaction at voltages above 4.5 V (vs. Li^+/Li). Accompanied formation of interphase and gaseous products have also been observed, attributed to

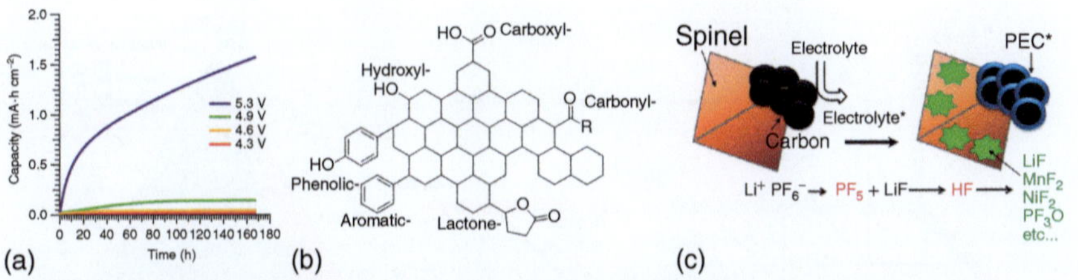

Figure 4.8 (a) Residual current (top) and capacity (bottom) of Li||Super P cells stored at different potentials. Source: Reproduced with permission Li et al. [61]. (b) Schematic drawing of possible functional groups at the carbon black periphery. Source: Reproduced with permission Qi et al. [63]. (c) A graph showing the parasitic reaction between electrolyte and carbon additive at high voltage. Source: Reproduced with permission Pieczonka et al. [64].

parasitic electrolyte oxidation, probably through self-discharge (the de-insertion of anions (PF_6^-)). The schematic graph of the anion insertion/de-insertion and parasitic electrolyte oxidation is shown in Figure 4.8c [64]. It can be concluded that the interface stability is an important parameter for long-term stability of novel high-voltage LIBs and LMBs.

Separators play a significant role in LIBs and LMBs, tunneling Li^+ between cathode and anode, and preventing short circuit [65]. The stability of conventional polymer-based separators is challenged at a high voltage above 4.7 V. Separators may suffer from pore blocking, porosity loss, and even mechanical fracture [66]. After long-term cycling, transition metal species can be detected on the separator in high-voltage Li||$LiNi_{0.5}Mn_{1.5}O_4$ cells. Chen and coworkers calculated the HOMO energy value of PE, PP separators and PVDF binder [67]. The results show that the PP separator is the easiest to be electrochemically oxidized among the three polymer materials. Coating ceramic layer is the common strategy to improve the interphase stability of separators, including at high-voltage environments.

In addition, few works have been focusing on the stability of binders and containers of LIBs and LMBs at high voltage. Polyvinylidene fluoride (PVdF), the most used binder for cathode electrodes, exhibits high electrochemical stability at high voltage [68]. Considering the chemical stability of C—F bonds, non-fluorinated binders probably suffer from oxidative decomposition [69]. Only a few kinds of cell containers have been studied at high voltage to understand their stability [70]. Surface passivation is believed to be helpful for practical applications of high-voltage LIBs and LMBs.

4.3 The Electrolyte Engineering for Various High-Voltage Cathodes

4.3.1 Nickel-Containing Layered Oxides

Being the dominant cathode material in LIB market for the past three decades, only about 145 mAh g^{-1} can be delivered by $LiCoO_2$ with an upper voltage limit of 4.3 V

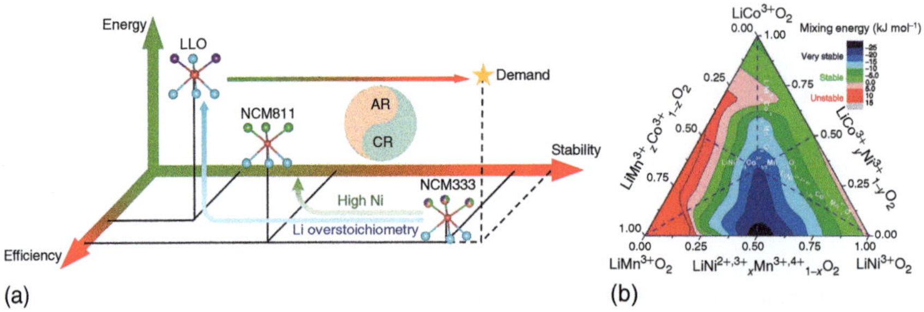

Figure 4.9 (a) A graph showing the relationship between energy density, efficiency, and stability for $LiNi_{1/3}Co_{1/3}Mn_{1/3}O_2$ cathode (NCM333), $LiNi_{0.8}Co_{0.1}Mn_{0.1}O_2$ cathode (NCM811), and lithium-rich layered oxide cathode. Source: Reproduced with permission Yin et al. [72]. (b) The ternary solid-solution system of Li-Ni-Co-Mn-O. Source: Reproduced with permission Liang et al. [73].

(vs. Li^+/Li). To meet the requirement of high energy from the power battery market, substitution of cobalt (Co) with other metal elements is an effective strategy, such as manganese (Mn), nickel (Ni), aluminum (Al). Until now, Ni-based layer oxides have attracted massive research attention and been enjoying great commercial success, for example, $LiNi_{1/3}Co_{1/3}Mn_{1/3}O_2$, $LiNi_{0.5}Co_{0.2}Mn_{0.3}O_2$, $LiNi_{0.8}Co_{0.1}Mn_{0.1}O_2$, and $LiNi_{0.8}Co_{0.15}Al_{0.05}O_2$ [71]. This class of cathode material shows higher reversible capacities of 160–200 mAh g^{-1} (depending on the Ni content) and higher upper cutoff voltages of 4.3–4.5 V than $LiCoO_2$. The relationship between energy density, efficiency, and stability of Ni-rich layer oxide cathodes is illustrated in Figure 4.9a [72]. As can be seen, increasing the Ni content can yield higher energy density because the Ni-rich cathode ($LiNi_{0.8}Co_{0.1}Mn_{0.1}O_2$ as example) has high reversible capacity of >200 mAh g^{-1} and the potential to operate at voltage >4.5 V. However, stability can be damaged by high Ni content, which is caused by two main reasons. First, among oxides with the same structural configuration, the surface oxygen of transition metals with greater electronegativity exhibits stronger nucleophilicity (Ni > Co > Mn) [73]. Due to the electrophilic properties of carbonate solvent molecules, Ni-rich oxides are more reactive toward carbonate-based electrolytes, compared to $LiCoO_2$ and other Ni-based oxides with low Ni content. Then, only a little of the nickel 3d band overlaps the top of the O 2p band, leading to greater accessible capacity of $LiNiO_2$ (about 240 mAh g^{-1}). However, due to similar ionic radii of Ni^{2+} and Li^+, the cation mixing easily occurs in Ni-rich oxides [74]. After Ni^{2+} migrates to Li slabs, Li^+ diffusion is impeded in the frameworks, causing the limited rate capability and the formation of surface residual Li species. In specific, during the delithiation process, multiple phase transition occurs. Finding the correlation between energy density (Ni content) and stability is the hot topic in the field in recent years. Figure 4.9b shows a ternary solid-solution system of Li–Ni–Co–Mn–O for which metal substitution can stabilize the layered framework of Ni-based cathode materials (Mn (4+) and Co (3+) for example) [75]. Al (3+) is another common substitute metal. After optimization, Ni-rich layered oxides in the commercial stage can show reversible capacity ranging from 160 to 200 mAh g^{-1}.

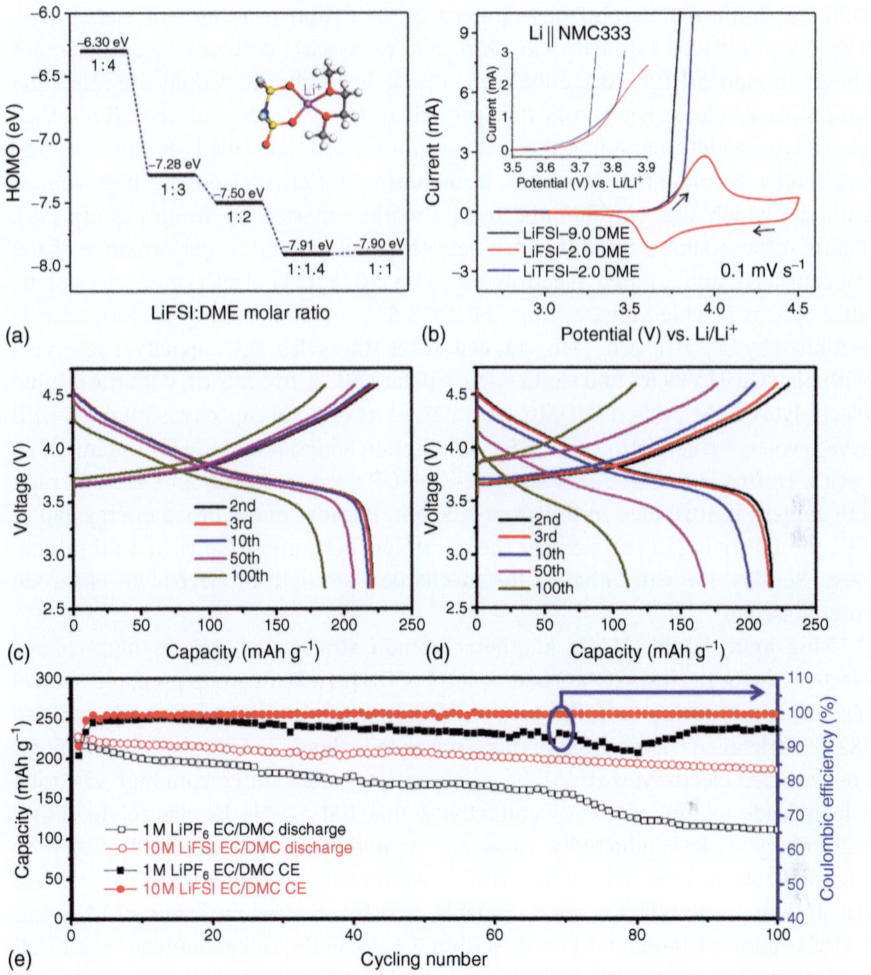

Figure 4.10 (a) The HOMO value of electrolyte with various salt (LiFSI) and solvent (DME) molar ratios. (b) CV curves of different ether-based electrolytes, for Li||$LiNi_{1/3}Co_{1/3}Mn_{1/3}O_2$ cell. Source: Reproduced with permission Ren et al. [78]. The inset shows the magnified region. Charge/discharge profiles of Li||$LiNi_{0.6}Co_{0.2}Mn_{0.2}O_2$ cell in (c) 10 M LiTFSI-EC/DMC electrolyte and (d) 1 M $LiPF_6$-EC/DMC electrolyte. (e) Cycling performance of Li||$LiNi_{0.6}Co_{0.2}Mn_{0.2}O_2$ cell in the two electrolytes, in the voltage range of 2.7–4.6 V. Source: Reproduced with permission Fan et al. [79].

It is well known that oxidation resistance of conventional carbonate-based electrolytes is limited, which cannot match that of Ni-rich oxides with a wide working voltage range [76, 77]. Therefore, strategies have been developed to solve this problem. Increasing the salt concentration can enhance oxidation resistance of electrolytes. Ren and coworkers studied the LiFSI-DME electrolyte with various concentrations [78]. As shown in Figure 4.10a, with the increasing concentration, the HOMO energy level of the electrolyte decreased and reached an ultimate value of −7.91 eV at the ratio of 1 : 1.4, suggesting the strongest oxidation resistance.

Different lithium salts also have impact on oxidation resistance of electrolytes. The CV results of Li||$LiNi_{1/3}Co_{1/3}Mn_{1/3}O_2$ cells with different electrolytes are shown in Figure 4.10b. As can be seen, although the concentrations are same, the LiTFSI-based electrolyte shows stronger oxidation resistance than the LiFSI-based electrolyte, which maintains the good operation of cells. This is because LiTFSI has higher F-content, helping the formation of F-rich CEI on the high-voltage cathode, which was corroborated by the works reported by Wang's group [79]. Figure 4.10c–e shows the difference between electrochemical performance of the concentrated and diluted electrolytes. With 10 M LiTFSI-EC/DMC electrolyte, although in a wide voltage range of 2.7–4.6 V, a stable cycling performance is exhibited by Li||$LiNi_{0.6}Co_{0.2}Mn_{0.2}O_2$ cell. After 100 cycles, 86% capacity is reserved, with high CE of >99.6% and slight voltage polarization. In contrast, with the diluted electrolyte of 1 M $LiPF_6$-EC/DMC, only 52% of its original capacity is retained with severe voltage polarization. The CE can maintain a high level only for the initial 20 cycles. During the subsequent 80 cycles, the CE decreases to about 95%. The high LiF content is attributed to this improvement. Because of the broad energy gap of LiF, the LiF-rich CEI can prevent the electrolyte decomposition with a thin thickness. Besides, LiF can enhance the mechanical strength of electrolyte–electrode interphase [80].

Using ionic liquid (IL) is another common strategy to prepare high-voltage electrolytes for the low HOMO energy level of ILs [81]. Dai's group prepared a novel electrolyte consisting of LiFSI salt, imidazolium-based IL, and NaTFSI additive [82]. The detailed composition of this electrolyte is shown in Figure 4.11a. Although concentrated electrolytes and IL-based electrolytes often suffer from high viscosity, which leads to limited ionic conductivity, this EM-5Li-Na IL electrolyte shows satisfactory ionic conductivity, especially at high temperatures, as displayed in Figure 4.10b. Besides, compared with conventional carbonate-based electrolytes, this IL-based electrolyte is nonflammable, which increases the safety of LIBs and LMBs (Figures 4.10d and 4.11c,d). Within 2.8–4.4 V, the cell assembled by Ni-rich oxide cathode with EM-5Li-Na IL shows good stability over 100 cycles at 0.5 C, as shown in Figure 4.11e.

In recent years, a new strategy has been developed, for example, bi-salt electrolytes. As shown in Figure 4.12a, at the same concentration, the bi-salt system of LiTFSI and LiFSI shows better oxidation resistance than the mono-salt system of LiFSI [83]. This is due to the novel electrolyte decomposition mechanism brought by the bi-salt system. In detail, on the cathode surface, the LiFSI decomposes firstly to form the LiF-containing interphase and the oxidation of LiTFSI is kinetically hindered. And the TFSI anions shield DME from direct contact with the cathode surface to stabilize the formed interphase. This novel bi-salt electrolyte can be applied to Ni-rich cathodes.

As mentioned earlier, the strong nucleophilicity of surface oxygen nickel oxides is one of the main reasons for electrolyte decomposition on Ni-rich cathodes. Therefore, efforts have been devoted to decrease surficial nucleophilicity through shielding the Ni at the surface atomic layer. Except surface coating [86–89], using film-forming additives can also achieve the goal [90–92]. For example, Liu et al.

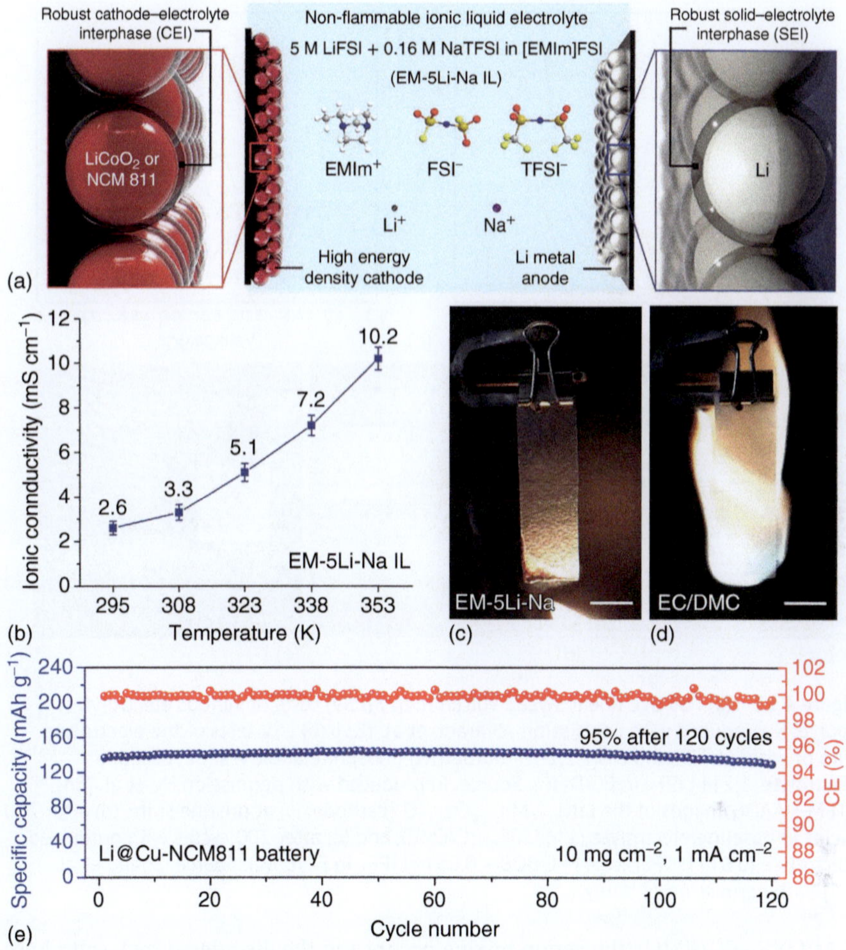

Figure 4.11 (a) Schematic graph of the cell configuration and electrolyte composition for the 1-ethyl-3-methyl-imidazolium-bis(fluorosulfonyl)imide (FSI)-based electrolyte (with 5 M LiFSI salt and 0.16 M NaTFSI additive), denoted as EM−5Li−Na IL. (b) Ionic conductivity of the as-prepared IL-based electrolyte at different temperatures. Digital graphs of the flammability tests for the (c) EM−5Li−Na IL electrolyte and (d) conventional organic electrolyte. (e) Cycling performance of the Li@Cu||$LiNi_{0.8}Co_{0.1}Mn_{0.1}O_2$ cell in EM−5Li−Na IL electrolyte within 2.8–4.4 V voltage range, at 0.5 C. Source: Sun et al. [82]/with permission of Wiley.

used triethlylphosphite (TEP) and tris(2,2,2-trifluoroethyl) phosphite (TTFP) as additives to improve the performance of Ni-rich cathodes [84]. As shown in Figure 4.12b, compared to the baseline electrolyte (BE), the additives are decomposed at lower potentials. As a result, the decomposition products can participate in the formation of the CEI and suppress electrolyte oxidation decomposition to increase CE. Interestingly, the change of electrolyte composition can not only adjust the formation of the CEI but also affect the phase transition of the Ni-rich cathode material. As can be seen from Figure 4.12c–e, lithiation and delithiation in BE

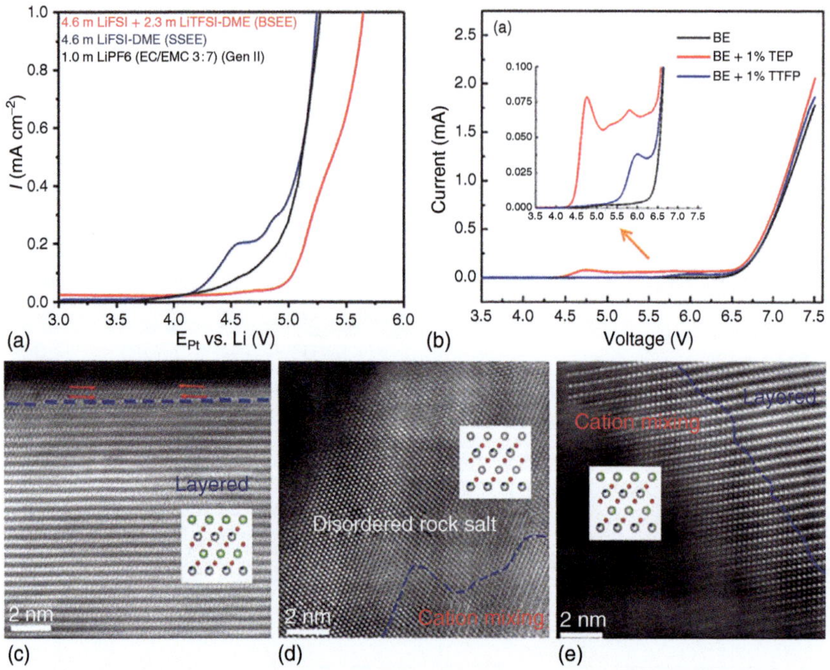

Figure 4.12 (a) Positive linear sweep voltammetry (LSV) tests of various electrolytes. Source: Reproduced with permission Alvarado et al. [83]. (b) LSV tests of the electrolyte with or without TEP and tris(2,2,2-trifluoroethyl) phosphite additives (BE: baseline electrolyte, 1.2 M $LiPF_6$ in EC/DMC). Source: Reproduced with permission He et al. [84]. STEM-HAADF images of the $LiNi_{0.76}Mn_{0.14}Co_{0.10}O_2$ cathode: (c) at pristine state, (d) after 200 cycles in baseline electrolyte (1 M $LiPF_6$-EC/EMC), and (e) after 200 cycles with optimized electrolyte (0.6 M LiTFSI + 0.4 M LiBOB + 0.05 M $LiPF_6$ in EC/EMC). Source: Zhang et al. [85]/with permission of Wiley.

(1 M $LiPF_6$-EC/EMC), the cation mixing occurs and the disordered rock-salt phase emerges [85]. By contrast, in the optimized electrolyte with the LiBOB additive, only a thin cation mixing layer emerges and no rock-salt phase can be detected.

4.3.2 $LiCoO_2$

In a common commercial application process, the upper cutoff voltage of $LiCoO_2$ cathode is often set to 4.2 V, corresponding to a capacity of about 140 mAh g^{-1}. With a high theoretical capacity of 274 mAh g^{-1}, there exists huge potential to increase reversible capacity and energy density of $LiCoO_2$. For example, if charging to 4.6 V, specific capacity of about 220 mAh g^{-1} and energy density of about 886 Wh kg^{-1} can be achieved [93–95]. In recent years, there have been massive research attention focused on high-voltage $LiCoO_2$ cathode materials. As mentioned, notorious phase transitions at high voltage hinder the application of $LiCoO_2$ as a high-voltage cathode. Cation substitution is the commonly used method to suppress those phase transitions, including Al, Mg, Ti, etc. The accompanying electrolyte issue should also be noted [96].

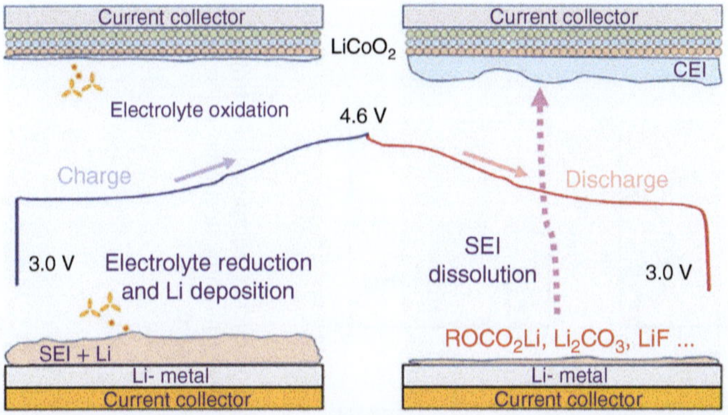

Figure 4.13 A schematic illustration of the formation process of CEI on $LiCoO_2$ cathode. Source: Reproduced with permission Lyu et al. [97].

Due to thermodynamically instability of organic species on high-voltage $LiCoO_2$ surface, the mechanisms of electrolyte decomposition and formation of CEI of this cathode are still complicated. Researchers have devoted their efforts in this field. For example, Li and coworkers put forward a novel mechanism of CEI formation on 4.6 V $LiCoO_2$ through species exchange with SEI. Figure 4.13 illustrates their assumption [97]. As can be seen, in a Li||$LiCoO_2$ cell during the charge process, a thick interphase forms on the Li anode for the electrolyte reduction and Li deposition. For the cathode side, only a thin CEI emerges. Some material exchange process occurs during this process. For instance, above 4.2 V, CO_2 generates from the $LiCoO_2$ side and migrates to the Li anode side. After contact with Li, Li_2CO_3 is produced. During the discharge process, the dissolution of SEI takes place. Some Li-based species migrate to the cathode surface and a thick CEI forms. This mechanism reminds us that SEI and CEI formation are not two independent issues [57, 98, 99].

For high-voltage $LiCoO_2$ cathodes, optimization of electrolytes is also important, which has impact on the interphase stability, cobalt ion dissolution, Li^+ diffusion in the electrode/electrolyte interface. In recent years, the electrolyte compatibility with high-voltage $LiCoO_2$ cathodes has largely enhanced. Many kinds of electrolyte additives have been reported to adjust the formation of CEI and protect $LiCoO_2$ cathodes [50, 100]. As a common film-forming additive, vinylene carbonate (VC) also has a positive influence on the electrochemical performance of $LiCoO_2$ [101]. As exhibited in Figure 4.14a,c, the electrons in different parts (in $LiCoO_2$ electrode and in electrolyte) have different electrochemical potentials. This difference causes the formation of a space-charge layer on the cathode. An electrical double layer also forms in the electrolyte, close to the cathode. Due to the low electrochemical potential of bulk $LiCoO_2$, Co^{3+} is reduced on the surface, with electrolyte oxidation. This leads to the formation of an irreversible interphase in the VC-free electrolyte. During the subsequent cycles, the above processes proceed repeatedly, resulting in increase in interfacial resistance and capacity decay. By contrast, in the VC-added electrolyte, the CV molecules are more easily oxidized on the $LiCoO_2$ surface than EC solvent

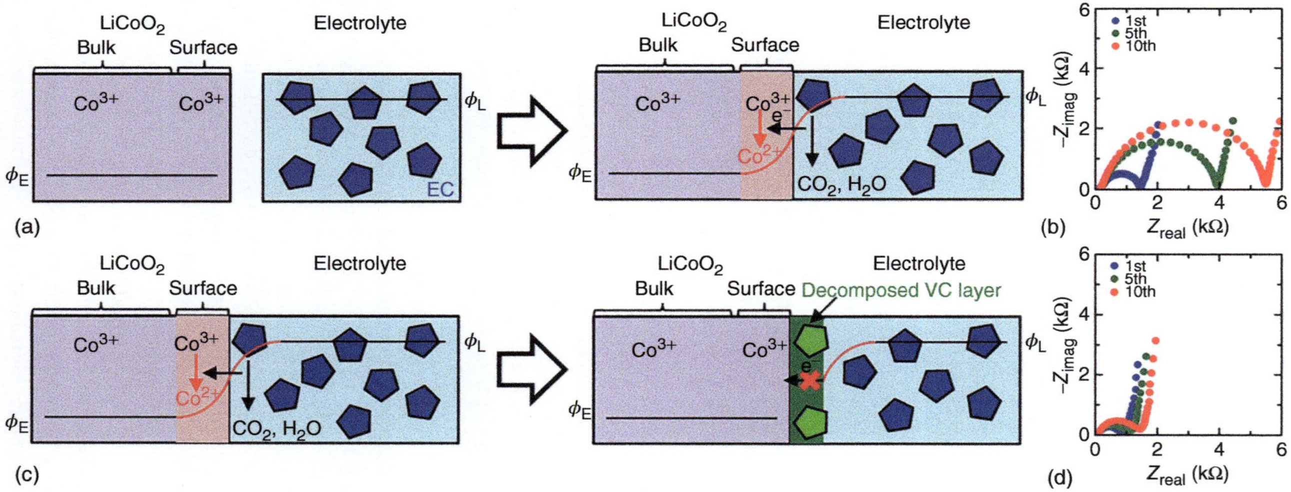

Figure 4.14 Schematic illustration showing the electronic structure of the $LiCoO_2$/electrolyte interface for (a) VC-free and (c) VC-containing electrolytes. ϕ_E: an electron in the electrode, ϕ_L: an electron in the electrolyte. Electrochemical impedance spectra (EIS) for the $LiCoO_2$ thin-film electrode: in the (b) VC-free and (d) VC-containing electrolytes, tested at 4.0 V (vs. Li^+/Li). Source: Reproduced with permission Takamatsu et al. [101].

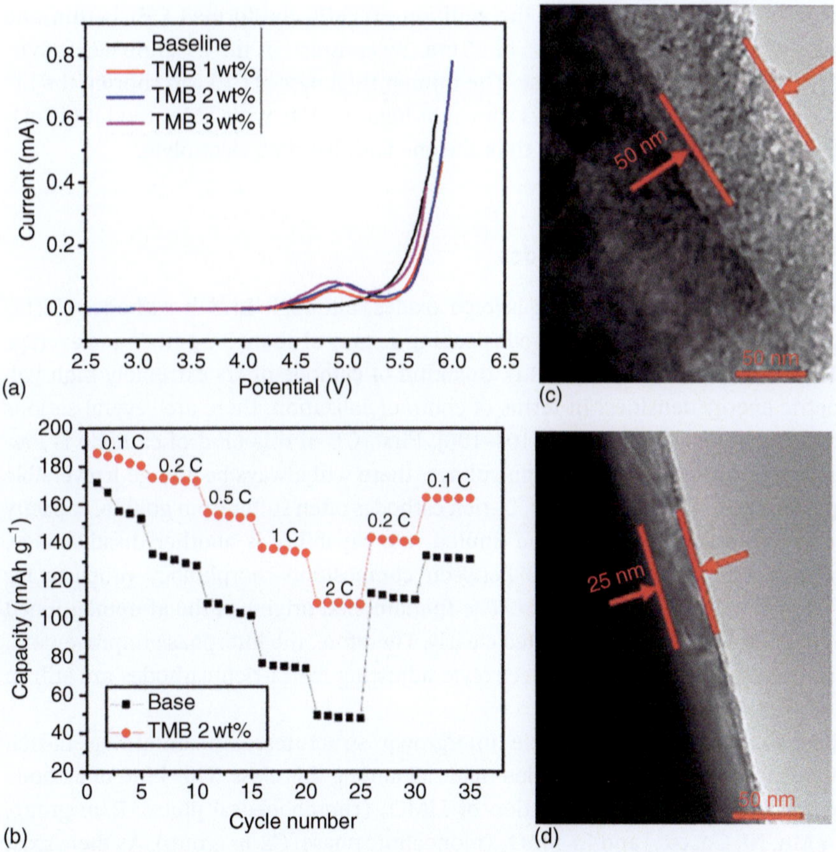

Figure 4.15 (a) Positive LSV profiles of the different electrolytes: baseline electrolyte (1.0 M $LiPF_6$-EC/DMC), and with various contents of trimethyl borate (TMB) additives. (b) The rate capabilities of Li||$LiCoO_2$ cells, with or without 2 wt% TMB, within a working voltage range of 2.5–4.5 V. TEM images of the CEI on $LiCoO_2$ cathode, formed in (c) baseline electrolyte and (d) 2 wt% TMB-containing electrolyte. Source: Liu et al. [103]/with permission of American Chemical Society.

molecules. The oxidized VC molecules will not be further oxidized, but polymerized. Therefore, the formed layer physically shields the $LiCoO_2$ with the electrolyte, suppressing the reduction of Co^{3+}. As can be seen from Figure 4.14b,d, with the VC-free electrolyte, the EIS of $LiCoO_2$ electrode continuously increases. The VC additive can suppress this phenomenon for the stable CEI.

Boron-containing additives are commonly used electrolyte additives, due to the stabilizing effect of boron oxides on the CEI [102]. For instance, Liu et al. used trimethyl borate (TMB) to stabilize the interphase of high-voltage $LiCoO_2$ [103]. As shown in Figure 4.15a, compared to the baseline electrolyte, the additive-containing electrolytes exhibit a lower oxidation potential at about 4.25 V, which is attributed to the film-forming reaction of TMB additive. Similar to LiF, boron oxides can also enhance the electronic isolation capability of CEI, leading a thinner thickness. As shown in Figure 4.15c,d, with the addition of TMB, the formed CEI is thin and dense, with a uniform thickness of 25 nm. By contrast, in the baseline electrolyte, the formed CEI is thick and loose. The thinner thickness of CEI can shorten the Li^+ diffusion distance. As a result, as shown in Figure 4.15b, with TMB, the Li||$LiCoO_2$ cell shows better rate capability than the one with baseline electrolyte.

4.3.3 Layered Li-Rich Cathodes

Compared to Li-stoichiometric layered oxides, although Li-rich cathodes ($Li(Li_x M_{1-x})O_2$) have a relatively electrode packing density of about 3.0 g cm^{-3} and average voltage of about 3.6 V (vs. Li^+/Li), this kind of cathode offers extremely high volumetric energy densities. In terms of commercialization, there are several serious problems for Li-rich cathodes [104–106]. First, CE of this kind of cathode is low. During the activation step at high voltage, there will always be a huge irreversible capacity, due to the Li and O loss. Li-rich cathodes often suffer from gradual capacity and voltage decline. Finally, the limited rate capability is another disadvantage. Unfortunately, the relationship between composition–morphology–property for Li-rich cathodes is quite complex. The fundamental origins of the aforementioned problems have not been elucidated clearly. Therefore, the interphase improvement, performance evaluation, and electrolyte adjusting for Li-rich cathodes are still in the preliminary stage [107, 108].

Because of highly compatible intergrown structures, Mn-containing Li-rich cathode is the most widely studied material among this class. This kind of cathode can be perceived as a combination of $LiMO_2$ (rhombohedral phase, $R3m$ group, M = Mn, Ni, Co, etc.) and Li_2MnO_3 (monoclinic phase, $C2/m$ group). As there exist multifold transition metal ions, the Mn-based Li-rich cathode undergoes a complex electrochemical reaction during the charging/discharging process. Figure 4.16a shows the electron energy levels of Mn^{4+}/Mn^{3+}, Ni^{4+}/Ni^{3+}, Ni^{3+}/Ni^{2+}, and Co^{4+}/Co^{3+} redox couples, which indicate the reaction orders of various metal ions [109]. The electrochemical reaction mechanism of $0.3Li_2MnO_3$–$0.7LiMn_{0.5}Ni_{0.5}O_2$ was studied by Thackeray et al., as an example [110]. When charging to 4.4 V, Li^+ will be extracted from the layered oxide with the redox couple of Ni^{4+}/Ni^{2+}, whereas the Li_2MnO_3 component is electrochemically inactive. The Mn-containing phase plays the role of Li^+ reservoir, meaning that Li^+ migrates to the Li^+-consumed layers. This process ensures the integrated structural stability of the Li-rich cathode. When charging to higher voltages, the Li loss and O release from Li_2MnO_3 phase will take place, namely activation process. The final product is metastable. During the discharging process, Li^+ can intercalate into the two phases and rock-salt $LiMnO_2$ and $LiMO_2$ phases form. Due to these complex electrochemical reaction mechanisms and interphase conditions, the electrolyte adjusting to match the Li-rich cathode is challenging.

The phase transition triggered by the surface Li loss and O release can cause not only the voltage and capacity decline but also lead to mechanical fracture of Li-rich cathode materials [111–113]. Jiang et al. tried to solve the mechanical fracture issue

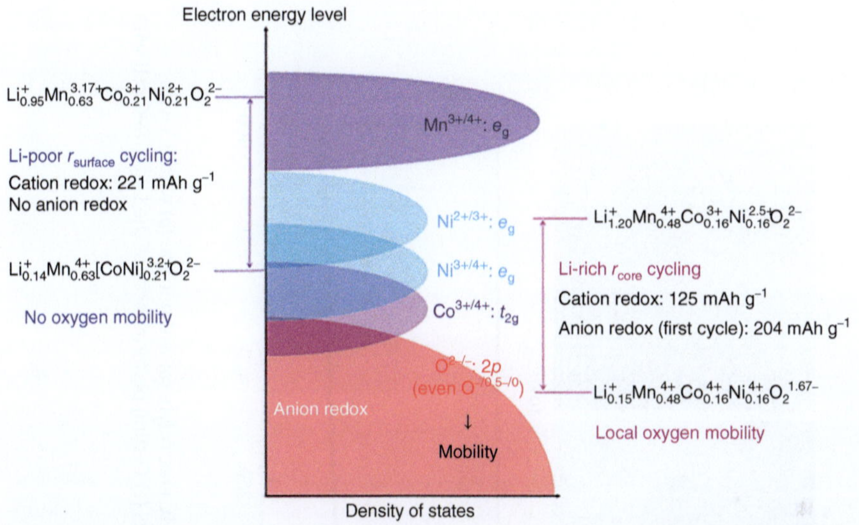

Figure 4.16 Redox reactions of transition metal cations and oxygen anions of the Li-poor surface and the Li-rich bulk. Source: Reproduced with permission Zhu et al. [109].

for $Li_{1.14}Ni_{0.133}Co_{0.133}Mn_{0.544}O_2$ (LRNMC) cathode by adding $LiPO_2F_2$ as the electrolyte additive [114]. As shown in Figure 4.17a, after 500 cycles at 3 C in the baseline electrolyte ($LiPF_6$-EC/DMC), the LRNMC particles show severe fragmentation. Benefited by the addition of $LiPO_2F_2$, mechanical fracture cannot be observed at the same condition, as displayed in Figure 4.17b. The TEM images of cycled LRNMC are shown in Figure 4.17c,d. After cycled in the baseline electrolyte, crystallinity of LRNMC cathode largely decreases. Importantly, a surficial spinel phase forms, with a thickness of about 8–15 nm. By contrast, after cycled with $LiPO_2F_2$ additive, the LRNMC maintains high crystallinity. The phase change zone is distributed on the particle surface, with a thin thickness of about 3 nm. It suggests the inhibition capability on phase transition of $LiPO_2F_2$ for Li-rich cathodes. The authors attributed this inhibition capability to the formation of dense CEI. Thus, the $LiPO_2F_2$ additive enhances surface and structural stability of LRNMC cathodes. As exhibited in Figure 4.17e, the cell with $LiPO_2F_2$ additive delivers better, long-term cycling performance than the one without this additive.

4.3.4 Other Cathode Materials

In addition to the above-discussed three kinds of cathode materials, spinel $LiNi_{0.5}Mn_{1.5}O_4$ and polyanionic compounds are the other classes of high-voltage cathode materials, which have been receiving research interests. Firstly, as a commercial cathode material, spinel lithium manganese oxide ($LiMn_2O_4$) has limited energy density, due to its low average voltage (<4 V) and available capacity (<145 mAh g^{-1}). To increase energy and power capabilities, manganese (Mn) ions are partially substituted with Ni ions and 5 V spinel $LiNi_{0.5}Mn_{1.5}O_4$ is obtained [115–118]. Such a big jump of the operating potential causes aggressive parasitic

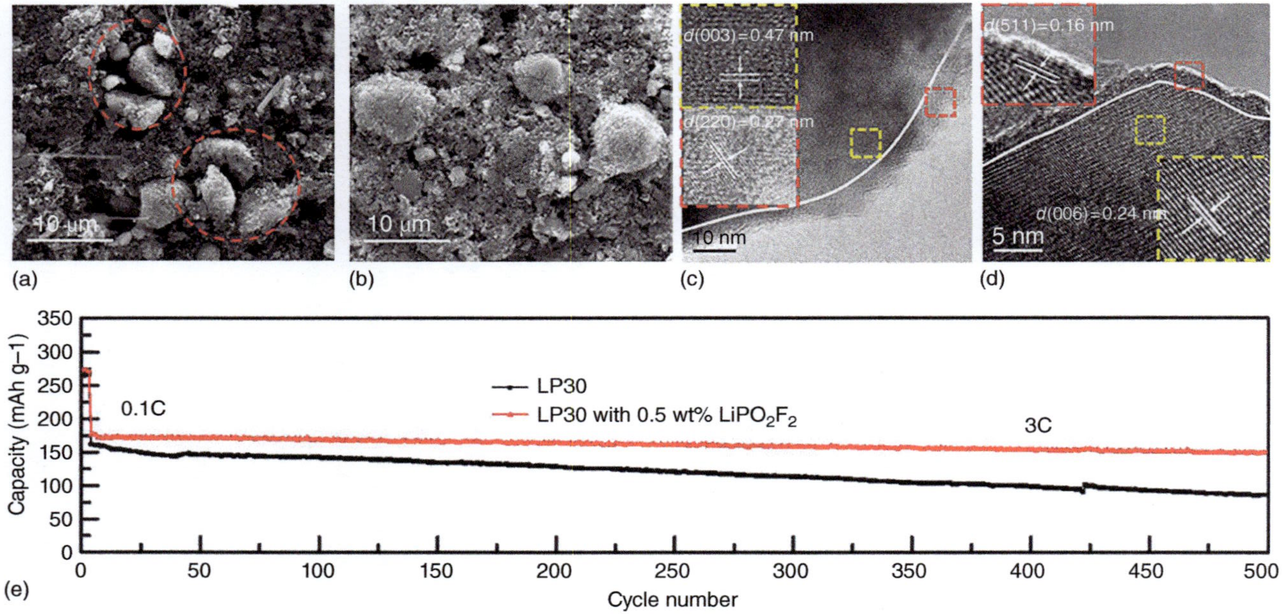

Figure 4.17 SEM images of the cycled $Li_{1.14}Ni_{0.133}Co_{0.133}Mn_{0.544}O_2$ (LRNMC) cathode in $LiPF_6$-EC/DMC electrolyte (a) without and (b) with 0.5 wt% $LiPO_2F_2$ additive. TEM images of the LRNMC cathode cycled (c) without and (d) with $LiPO_2F_2$ additive. (e) Cycling performances of the Li||LRNMC cell with and without $LiPO_2F_2$ additive. Source: Jiang et al. [114]/with permission of Elsevier.

electrolyte reactions in conventional carbonate-based electrolytes [119]. To solve this issue, some novel electrolyte systems have been developed. For instance, sulfone generally has a lower HOMO value than that of carbonate, meaning a higher oxidation potential. Amine's group use EMS as solvent to prepare the electrolyte for $LiNi_{0.5}Mn_{1.5}O_4$, as shown in Figure 4.18 [120]. Cycled within the voltage range of 3.5–4.9 V, the graphite|| $LiNi_{0.5}Mn_{1.5}O_4$ cell with the EMS-based electrolyte shows a capacity retention rate of 80.8% over 100 cycles.

The electrolyte additives for $LiNi_{0.5}Mn_{1.5}O_2$ cathode have also been developed and reported. For example, Liu and coworkers used ethoxy(pentafluoro) cyclotriphosphazene (PFPN) to adjust the interphase and electrochemical performance of $LiNi_{0.5}Mn_{1.5}O_2$ cathode [121]. The mechanism of this kind of additive is shown in Figure 4.19a. On the cathode particles surface, due to the nucleophilic attack by the surface O atom in $LiNi_{0.5}Mn_{1.5}O_2$, the ring of PFPN opens and polymerization occurs. Almost at the same time, the oxidative decomposition of phosphazene takes place to produce inorganic products of LiN_xO_y and $Li_xPO_yF_z$. As shown in Figure 4.19b,c, without the PFPN additive, the formed CEI has a nonuniform thickness of 0.9–4.5 nm. And a low crystalline region can be observed, implying the phase transition. By contrast, with the PFPN additive, the formed organic/inorganic hybrid CEI has a thin and uniform thickness of about 1.3 nm. The structure of $LiNi_{0.5}Mn_{1.5}O_2$ cathode is preserved well. (The interplanar spacing of 0.49 nm corresponds to the [111] planes of $LiNi_{0.5}Mn_{1.5}O_2$.) Benefited by the thin CEI and high structural stability brought about by the PFPN additive, the high mass loading $LiNi_{0.5}Mn_{1.5}O_2$ electrode shows a good cycle life over 100 cycles, as shown in Figure 4.19d.

Polyanionic compounds are a large class of Li^+ host framework cathodes. Among various polyanionic compounds, lithium iron phosphate ($LiFePO_4$) is the most famous one for its successful commercialization [122]. However, reversible capacity and operating voltage of $LiFePO_4$ are relatively low (about 160 mAh g^{-1} and 3.4 V, respectively), which limit its energy density. Fortunately, abundant compositions and rich crystal chemistries of these compounds bring feasible designability for their capacity and operating voltage. In recent years, some high-voltage polyanionic compound cathodes have been reported, including $LiCoPO_4$, and $LiNiPO_4$ [123–125]. However, compared to the research about layered oxide cathodes, studies focused on polyanionic compound cathodes are less, maybe due to their low powder densities and intrinsic-limited electrochemical kinetics. Considering their safety and cost advantages, it is worth to continue improving this class of high-voltage cathodes.

To match electrolytes with polyanionic compound cathodes is still challenging. Fan et al. prepared a novel electrolyte for $LiCoPO_4$, composed of $LiPF_6$ fluoroethylene carbonate/3,3,3-fluoroethylmethyl carbonate/1,1,2,2-tetrafluoroethyl-2′,2′,2′-trifluoroethyl ether (FEC:FEMC:HFE) [126]. The key factor of this electrolyte is high fluorine content. As shown in Figure 4.20a, the defluorination reaction of Li^+-solvent and Li^+-PF_6 pairs will occur within 1.5–2.5 V to form LiF, as predicted by quantum chemistry calculations. As mentioned earlier, the LiF-rich CEI has better electron shielding capability and mechanical strength. The formed Li-rich

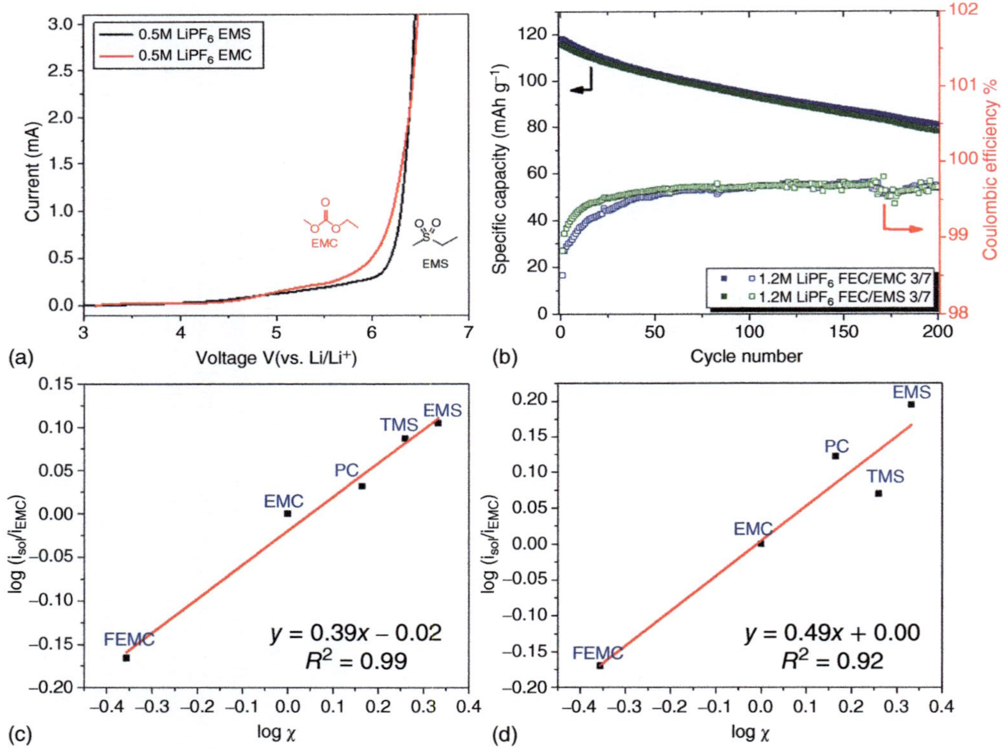

Figure 4.18 (a) Positive linear sweep for the two electrolytes of 0.5 M $LiPF_6$-EMC and 0.5 M $LiPF_6$-EMS. (b) Cycling performances for the graphite||$LiNi_{0.5}Mn_{1.5}O_2$ cell with the electrolytes of 1.2 M $LiPF_6$-FEC/EMC and 1.2 M $LiPF_6$-FEC/EMS. Chi(χ)-plots ($\log(i_{sol}/i_{EMC})$ against $\log\chi$) at (c) 4.6 V and (d) 4.3 V. Source: Reproduced with permission Su et al. [120].

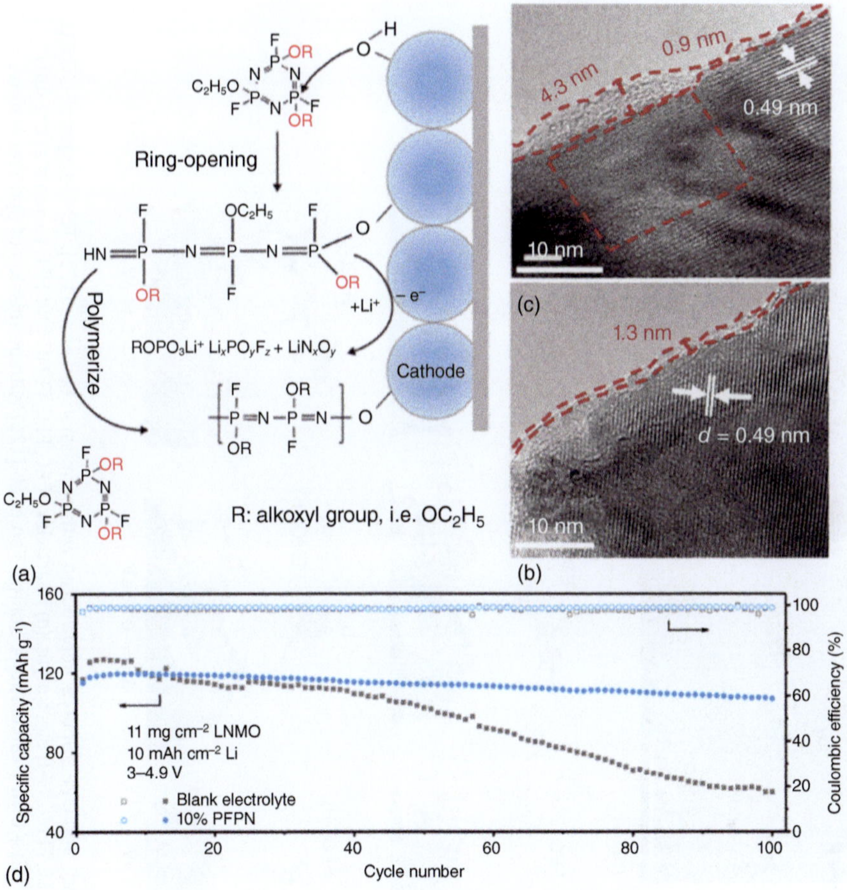

Figure 4.19 (a) Illustration of the possible decomposition mechanism of ethoxy(pentafluoro)cyclotriphosphazene (PFPN) additive on the cathode side. TEM images of the $LiNi_{0.5}Mn_{1.5}O_2$ cathode, cycled in the electrolyte of $LiPF_6$-EC/DEC, (b) with and (c) without PFPN additive. (d) Cycling performances of the Li||$LiNi_{0.5}Mn_{1.5}O_2$ cell with and without PFPN additive. Source: Liu et al. [121]/with permission of Elsevier.

CEI protects the $LiCoPO_4$ cathode from aggressive phase transition. As can be seen from Figure 4.20b,c, after some cycles, an amorphous layer formed on the cathode surface and the $LiCoPO_4$ particles maintain high crystallinity. The detailed decomposition reaction mechanisms of solvent molecules are further investigated by density functional theory (DFT) calculations. As shown in Figure 4.20d, for the $CoPO_4$ (the fully charged state of $LiCoPO_4$), the H transferring reaction occurs. The H of EC molecules is most preferred to transfer to react with surface O of $CoPO_4$, whereas the H of HFE is the most difficult to participate in this reaction. The residual part of EC ($EC\cdot_{(-H)}$ free radical) is easy to evolve CO_2, or to participate in polymerization. However, the products of the two reactions are detrimental to the CEI. By contrast, the $FEC\cdot_{(-H)}$ free radicals are a high barrier to the ring opening reaction, which are prone to react with $FEC\cdot_{(-H)}$ and $HFE\cdot_{(-H)}$ radicals to form the

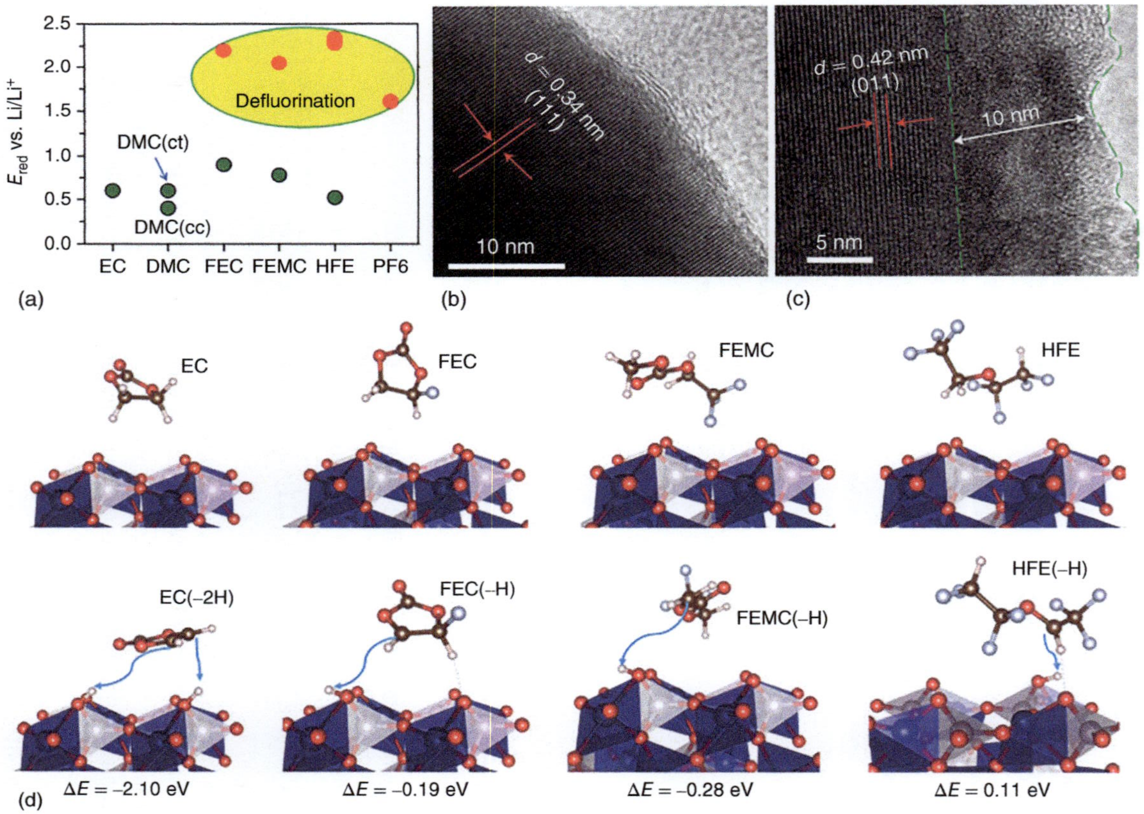

Figure 4.20 (a) Calculated reduction potentials of Li^+-(solvent)$_n$ and $LiPF_6$. Source: Reproduced with permission Fan et al. [125]. TEM images of $LiCoPO_4$ (LCP) cathode (b) at pristine state and (c) cycled in 1 M $LiPF_6$-FEC/FEMC/HFE electrolyte. (d) Reactivity of EC, FEC, FEMC, and HFE solvents at the fully charged $CoPO_4$ (010) surface from Perdew–Burke–Enzerhoff (PBE) + U DFT calculations. Source: Fan et al. [126]/with permission of Springer Nature.

F-rich CEI. In addition, fluorinated radicals perhaps bind to surface O of $CoPO_4$, preventing transition metal dissolution.

4.4 Conclusions

Batteries with higher energy density have been the research focus, which drive the progress of electric vehicles. Development of high-voltage cathodes, including Ni-rich oxides, high-voltage $LiCoO_2$, Li-rich oxides, spinel $LiNi_{0.5}Mn_{1.5}O_2$, and polyanionic compounds, not only brings the application progress but also challenges to electrolytes. Electrolyte oxidation decomposition, surface structure transformation, transition metal ion dissolution, mechanical fracture, and impedance growth disrupt high-voltage batteries. All these issues are related to electrolytes. Therefore, electrolyte regulation and optimization play a decisive role in meeting energy density requirements of power batteries for electric vehicles. The formation mechanism of CEI, electrolyte decomposition, parasitic reaction, structure reconstruction of high-voltage cathode materials, and stability of other cell components at high voltage were discussed in this chapter. Besides, the strategies of electrolyte optimization for various high-voltage cathode materials were also illustrated.

References

1 Goodenough, J.B. and Park, K.-S. (2013). *Journal of the American Chemical Society* 135: 1167–1176.
2 Dunn, B., Kamath, H., and Tarascon, J.-M. (2011). *Science* 334: 928–935.
3 Tarascon, J.M. and Armand, M. (2001). *Nature* 414: 359–367.
4 Manthiram, A. (2011). *Journal of Physical Chemistry Letters* 2: 176–184.
5 Jiao, S.H., Ren, X.D., Cao, R.G. et al. (2018). *Nature Energy* 3: 739–746.
6 Cheng, X.B., Zhang, R., Zhao, C.Z., and Zhang, Q. (2017). *Chemical Reviews* 117: 10403–10473.
7 Lu, J., Li, L., Park, J.-B. et al. (2014). *Chemical Reviews* 114: 5611–5640.
8 Whittingham, M.S. (1976). *Science* 192: 1126.
9 Whittingham, M.S. (1976). *Journal of the Electrochemical Society* 123: 315–320.
10 Mizushima, K., Jones, P.C., Wiseman, P.J., and Goodenough, J.B. (1980). *Materials Research Bulletin* 15: 783–789.
11 Wang, J.J. and Sun, X.L. (2015). *Energy & Environmental Science* 8: 1110–1138.
12 Xu, G.J., Liu, Z.H., Zhang, C.J. et al. (2015). *Journal of Materials Chemistry A* 3: 4092–4123.
13 Wang, X., Wang, X., and Lu, Y. (2019). *Industrial and Engineering Chemistry Research* 58: 10119–10139.
14 Kim, J., Cho, H., Jeong, H.Y. et al. (2017). *Advanced Energy Materials* 7: 8.
15 Choi, J.W. and Aurbach, D. (2016). *Nature Reviews Materials* 1: 16013.
16 Croguennec, L. and Palacin, M.R. (2015). *Journal of the American Chemical Society* 137: 3140–3156.

17 Huang, Y.Y., Zhu, Y.C., Fu, H.Y. et al. (2021). *Angewandte Chemie International Edition* 60: 4682–4688.

18 Hwang, S., Kim, D.-H., Shin, J.H. et al. (2018). *The Journal of Physical Chemistry C* 122: 19438–19446.

19 Song, K.M., Liu, C.T., Mi, L.W. et al. (2021). *Small* 17: 1903194.

20 Chen, X., Li, H.X., Yan, Z.H. et al. (2019). *Science China Materials* 62: 1515–1536.

21 Wu, Q., Mao, S., Wang, Z. et al. (2020). *Nano Select* 1: 111–134.

22 Mao, S.L., Wu, Q., Ma, F.Y. et al. (2021). *Chemical Communications* 57: 840–858.

23 Li, W.D., Song, B.H., and Manthiram, A. (2017). *Chemical Society Reviews* 46: 3006–3059.

24 Yang, H., Li, J., Sun, Z. et al. (2020). *Energy Storage Materials* 30: 113–129.

25 Zhang, H., Liu, X., Li, H.H. et al. (2021). *Angewandte Chemie International Edition* 60: 598–616.

26 Palacın, M.R. and de Guibert, A. (2016). *Science* 351: 1253292.

27 Verma, P., Maire, P., and Nova'k, P. (2010). *Electrochimica Acta* 55: 6332–6341.

28 Zheng, J., Lochala, J.A., Kwok, A. et al. (2017). *Advanced Science* 4: 1700032.

29 Tikekar, M.D., Choudhury, S., Tu, Z.Y., and Archer, L.A. (2016). *National Energy* 1: 1–7.

30 Liu, W., Oh, P., Liu, X. et al. (2015). *Angewandte Chemie International Edition* 54: 4440–4457.

31 Kim, T., Song, W.T., Son, D.-Y. et al. (2019). *Journal of Materials Chemistry A* 7: 2942–2964.

32 Zhong, Q., Bonakdarpour, A., Zhang, M. et al. (1997). *Journal of the Electrochemical Society* 144: 205–213.

33 Gauthier, M., Carney, T.J., Grimaud, A. et al. (2015). *The Journal Physical Chemistry Letters* 6: 4653–4672.

34 Xu, G.J., Liu, Z.H., Zhang, C.J. et al. (2015). *Journal of Materials Chemistry A* 3: 4092–4123.

35 Levi, M.D., Salitra, G., Markovsky, B. et al. (1999). *Journal of the Electrochemical Society* 146: 1279–1289.

36 Zhang, L., Tarascon, J.-M., Sougrati, M.T. et al. (2015). *Journal of Materials Chemistry A* 3: 16988–16997.

37 Li, G., Huang, Z., Zuo, Z. et al. (2014). *Materials and Technology* 29: A77–A81.

38 Yang, L., Ravdel, B., and Lucht, B.L. (2010). *Electrochemical and Solid-State Letters* 13: A95–A97.

39 Browning, J.F., Baggetto, L., Jungjohann, K.L. et al. (2014). *ACS Applied Materials and Interface* 6: 18569–18576.

40 Chen, Y., Zhao, W., Zhang, Q. et al. (2020). *Advanced Functional Materials* 30: 2000396.

41 Li, W.D., Dolocan, A., Oh, P. et al. (2017). *Nature Communications* 8: 14589.

42 Tripathi, A.M., Su, W.N., and Hwang, B.J. (2018). *Chemical Society Reviews* 47: 736–851.

43 Kasnatscheew, J., Evertz, M., Streipert, B. et al. (2016). *Physical Chemistry Chemical Physics* 18: 3956–3965.
44 Wu, K., Yang, J., Liu, Y. et al. (2013). *Journal of Power Sources* 237: 285–290.
45 Xiong, D.J., Petibon, R., Nie, M. et al. (2016). *Journal of the Electrochemical Society* 163: A546–A551.
46 Jung, S.-K., Gwon, H., Hong, J. et al. (2014). *Advanced Energy Materials* 4: 1300787.
47 Hwang, S., Chang, W., Kim, S.M. et al. (2014). *Chemistry of Materials* 26: 1084–1092.
48 Yan, P.F., Nie, A.M., Zheng, J.M. et al. (2015). *Nano Letters* 15: 514–522.
49 Oh, P., Ko, M., Myeong, S. et al. (2014). *Advanced Energy Materials* 4: 1400631.
50 Zhu, Y., Luo, X., Zhi, H. et al. (2017). *ACS Applied Materials & Interfaces* 9: 12021–12034.
51 Wu, H., Chan, G., Choi, J.W. et al. (2012). *Nature Nanotechnology* 7: 309–314.
52 Reddy, M.V., Yu, T., Sow, C.H. et al. (2007). *Advanced Functional Materials* 17: 2792–2799.
53 Wan, J., Zhu, J., Xiang, Y. et al. (2021). *Journal of Energy Chemistry* 54: 786–794.
54 Lee, E.-J., Chen, Z., Noh, H.-J. et al. (2014). *Nano Letters* 14: 4873–4880.
55 Sun, H.H., Ryu, H.-H., Kim, U.-H. et al. (2020). *ACS Energy Letters* 5: 1136–1146.
56 Mao, M., Huang, B., Li, Q. et al. (2020). *Nano Energy* 78: 105282.
57 Pieczonka, N.P.W., Liu, Z., Lu, P. et al. (2013). *The Journal of Physical Chemistry C* 117: 15947–15957.
58 Zhan, C., Wu, T.P., Lu, J., and Amine, K. (2018). *Energy & Environmental Science* 11: 243–257.
59 Lyu, Y., Wu, X., Wang, K. et al. (2020). *Advanced Energy Materials*: 2000982. https://doi.org/10.1002/aenm.202000982.
60 Kim, J.-H., Pieczonka, N.P.W., Li, Z. et al. (2013). *Electrochimica Acta* 90: 556–562.
61 Li, X., Chen, Y., Nguyen, C.C. et al. (2014). *Journal of the Electrochemical Society* 161: A576–A582.
62 Matsumoto, K., Inoue, K., Nakahara, K. et al. (2013). *Journal of Power Sources* 231: 234–238.
63 Qi, X., Blizanac, B., DuPasquier, A. et al. (2015). *Journal of the Electrochemical Society* 162: A339–A343.
64 Pieczonka, N.P.W., Yang, L., Balogh, M.P. et al. (2013). *The Journal of Physical Chemistry C* 117: 22603–22612.
65 Lu, Y., Tu, Z., and Archer, L.A. (2014). *Nature Materials* 13: 961–969.
66 Kostecki, R., Norin, L., Song, X., and McLarnon, F. (2004). *Journal of the Electrochemical Society* 151: A522–A526.
67 Chen, X., Xu, W., Xiao, J. et al. (2012). *Journal of Power Sources* 213: 160–168.
68 Yoon, T., Park, S., Mun, J. et al. (2012). *Journal of Power Sources* 215: 312–316.
69 Li, J., Klöpsch, R., Nowak, S. et al. (2011). *Journal of Power Sources* 196: 7687–7691.

70 Sinha, N.N., Burns, J.C., Sanderson, R.J., and Dahn, J. (2011). *Journal of the Electrochemical Society* 158: A1400–A1403.

71 Myung, S.-T., Maglia, F., Park, K.-J. et al. (2017). *ACS Energy Letters* 2: 196–223.

72 Yin, C., Wan, L., Qiu, B. et al. (2021). *Energy Storage Materials* 35: 388–399.

73 Aurbach, D., Markovsky, B., Salitra, G. et al. (2007). *Journal of Power Sources* 165: 491–499.

74 Dahn, J.R., von Sacken, U., Juzkow, M.W., and Al-Janaby, H. (1991). *Journal of the Electrochemical Society* 138: 2207–2211.

75 Liang, C., Kong, F., Longo, R.C. et al. (2016). *The Journal of Physical Chemistry C* 120: 6383–6393.

76 Wang, Q., Yao, Z., Zhao, C. et al. (2020). *Nature Communications* 11: 4188.

77 Peng, Z., Cao, X., Gao, P. et al. (2020). *Advanced Functional Materials* 30: 2001285.

78 Ren, X., Zou, L., Jiao, S. et al. (2019). *ACS Energy Letters* 4: 896–902.

79 Fan, X.L., Chen, L., Ji, X. et al. (2018). *Chem* 4: 174–185.

80 Suo, L., Xue, W., Gobet, M. et al. (2018). *National Academy Science USA* 115: 1156.

81 Yang, Q.W., Zhang, Z.Q., Sun, X.G. et al. (2018). *Chemical Society Reviews* 47: 2020–2064.

82 Sun, H., Zhu, G.Z., Zhu, Y.M. et al. (2020). *Advanced Materials* 32: 2001741.

83 Alvarado, J., Schroeder, M.A., Pollard, T.P. et al. (2019). *Energy and Environmental Science* 12: 780–794.

84 He, M., Su, C.C., Peebles, C. et al. (2016). *ACS Applied Materials and Interface* 8: 11450–11458.

85 Zhao, W., Zheng, J., Zou, L. et al. (2018). *Advanced Energy Materials*: 1800297.

86 Myung, S.-T., Izumi, K., Komaba, S. et al. (2005). *Chemistry of Materials* 17: 3695–3704.

87 Wise, A.M., Ban, C., Weker, J.N. et al. (2015). *Chemistry of Materials* 27: 6146–6154.

88 Yoon, W.-S., Nam, K.-W., Jang, D. et al. (2012). *Journal of Power Sources* 217: 128–134.

89 Park, M.-H., Noh, M., Lee, S. et al. (2014). *Nano Letters* 14: 4083–4089.

90 Zheng, Y., Fang, W., Zheng, H. et al. (2019). *Journal of the Electrochemical Society* 166: A3222–A3227.

91 Wu, S., Lin, Y., Xing, L. et al. (2019). *ACS Applied Materials and Interfaces* 11: 17940–17951.

92 Yoo, D.-J., Kim, K.J., and Choi, J.W. (2018). *Advanced Energy Materials* 8: 1702744.

93 Zhang, J.-N., Li, Q., Ouyang, C. et al. (2019). *Nature Energy* 4: 594–603.

94 Qian, J., Liu, L., Yang, J. et al. (2018). *Nature Communications* 9: 4918.

95 Wang, L., Ma, J., Wang, C. et al. (2019). *Advancement of Science* 6: 1900355.

96 Yu, Y., Kong, W., Yang, W. et al. (2021). *Journal of Physical Chemistry C* 125: 2364–2372.

97 Zhang, J.-N., Li, Q., Wang, Y. et al. (2018). *Energy Storage Materials* 14: 1.

98 Lu, D., Xu, M., Zhou, L. et al. (2013). *Journal of the Electrochemical Society* 160: A3138–A3143.
99 Jarry, A., Gottis, S., Yu, Y.-S. et al. (2015). *Journal of the American Chemical Society* 137: 3533–3539.
100 Tang, C., Ren, Y., Chen, Z., and Ding, J. (2019). *Ionics* 25: 1007–1014.
101 Takamatsu, D., Orikasa, Y., Mori, S. et al. (2015). *The Journal of Physical Chemistry C* 119: 9791–9797.
102 Zuo, X., Fan, C., Liu, J. et al. (2013). *Journal of Power Sources* 229: 308–312.
103 Liu, Q., Yang, G., Liu, S. et al. (2019). *ACS Applied Materials & Interfaces* 11: 17435–17443.
104 Xiang, X.D., Knight, J.C., Li, W.S., and Manthiram, A. (2014). *The Journal of Physical Chemistry C* 118: 21826–21833.
105 Ren, W.F., Zhao, Y.J., Hu, X.S., and Xia, M.H. (2016). *Electrochimica Acta* 191: 491–499.
106 Lengyel, M., Atlas, G., Elhassid, D. et al. (2014). *Journal of Power Sources* 262: 286–296.
107 Hu, S., Pillai, A.S., Liang, G. et al. (2019). *Electrochemical Energy Reviews* 2: 277–311.
108 Birrozzi, A., Laszczynski, N., Hekmatfar, M. et al. (2016). *Journal of Power Sources* 325: 525–533.
109 Zhu, Z., Yu, D., Yang, Y. et al. (2019). *Nature Energy* 4: 1049–1058.
110 Thackeray, M.M., Kang, S.H., Johnson, C.S. et al. (2006). *Electrochemistry Communications* 8: 1531–1538.
111 Jarvis, K.A., Deng, Z.Q., Allard, L.F. et al. (2011). *Chemistry of Materials* 23: 3614–3621.
112 Bareno, J., Balasubramanian, M., Kang, S.H. et al. (2011). *Chemistry of Materials* 23: 2039–2050.
113 Shukla, A.K., Ramasse, Q.M., Ophus, C. et al. (2015). *Nature Communications* 6: 8711.
114 Jiang, B., Li, J., Luo, B. et al. (2021). *Journal of Energy Chemistry* 60: 564–571.
115 Xu, M., Zhou, L., Dong, Y. et al. (2016). *Energy & Environmental Science* 9: 1308–1319.
116 Kim, J.-H., Pieczonka, N.P.W., Lu, P. et al. (2015). *Advanced Materials Interfaces* 2: 1500109.
117 Dudney, N.J. (2005). *Materials Science & Engineering, B: Advanced Functional Solid-State Materials* 116: 245–249.
118 Li, J., Ma, C., Chi, M. et al. (2015). *Advanced Energy Materials* 5: 1401408.
119 Wang, J., Yamada, Y., Sodeyama, K. et al. (2016). *Nature Communications* 7: 12032.
120 Su, C.-C., He, M., Amine, R. et al. (2021). *Nano Energy* 83: 105843.
121 Liu, Q., Chen, Z., Liu, Y. et al. (2021). *Energy Storage Materials* 37: 521–529.
122 Yamada, A., Chung, S.C., and Hinokuma, K. (2001). *Journal of the Electrochemical Society* 148: A224–A229.
123 Ben Bechir, M., Ben Rhaiem, A., and Guidara, K. (2014). *Bulletin of Materials Science* 37: 473–480.

124 Amine, K., Yasuda, H., and Yamachi, M. (2000). *Electrochemical and Solid-State Letters* 3: 178–179.

125 Delp, S.A., Borodin, O., Olguin, M. et al. (2016). *Electrochimica Acta* 209: 498–510.

126 Fan, X., Chen, L., Borodin, O. et al. (2018). *Nature Nanotechnology* 13: 715–722.

5

Extreme Temperature Electrolytes

5.1 Low-Temperature Electrolytes

Conventional lithium-based batteries exhibit outstanding performance when used at ambient temperature. As shown in Figure 5.1, many applications have to be used at low temperature [1–4]. Take electric vehicles as example, poor performance such as energy and capacity loss at subzero temperature in winter is a significant limitation for mass promotion compared to traditional internal combustion vehicles [5]. Like 18 650 Li–ion cells, the power density decreases from ~800 to <10 $W\,l^{-1}$ and the energy density decreases from 100 to ~5 $Wh\,l^{-1}$ when the temperature decreases from 25 to −40 °C [6]. What's more, spaceflight such as Mars Exploration requires the batteries operating at ultralow-temperature even below −80 °C [7]. Most of the commercial electrolytes solidified at such low a temperature due to the high freezing point cause the failure of battery [8]. Therefore, it is necessary to explore electrolytes that can still work at low temperature to meet the needs of special batteries.

5.1.1 The Limitations of Battery Performance at Low Temperature

As temperature decreases, the kinetic rate of chemical and electrochemical reactions is slower than at ambient temperature [9, 10]. Meanwhile, the electrolyte will become more viscous and even solidify, resulting in a rapid decline in conductivity [11]. The power and energy densities of cells decrease sharply when the temperature is below −20 °C [12, 13]. The physical properties of some solvents are listed in Table 5.1. Ethylene carbonate (EC) is often regarded as an important component of solvents in commercial electrolytes with high dielectric permittivity of 89.6 for better dissolution of the Li salts, but its freezing point is as high as 36 °C, which implies the difficulty using it as a low-temperature electrolyte [15, 16]. Straight-chain solvents such as dimethyl carbonate (DMC) and ethylmethyl carbonate (EMC) have relatively low viscosity and good electrochemical stability, indicating their practicability [17]. In addition, introducing aliphatic esters as cosolvents into electrolytes to further decrease the freezing point is also a feasible method [18]. The rapid reduction in conductivity at low temperature is also an important issue. While the conductivity

Liquid Electrolyte Chemistry for Lithium Metal Batteries: Design, Mechanisms, Strategies,
First Edition. Jianmin Ma.

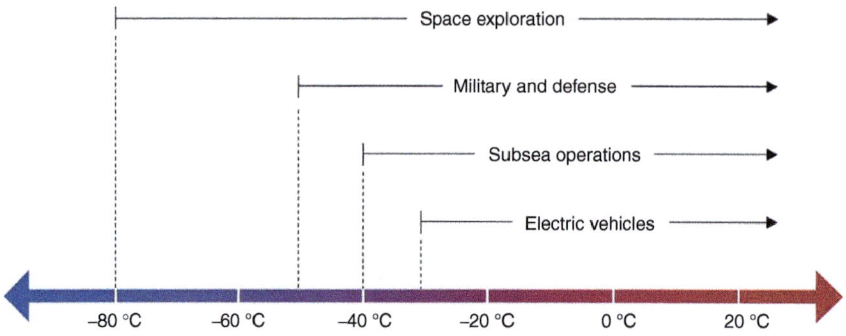

Figure 5.1 The operating temperature range of different battery systems at low temperature. Source: Abhay and Manthiram [1].

Table 5.1 Physical properties of some solvents.

Solvent	ε	η (cP)	fp (°C)	bp (°C)
Toluene (tol)	2.379	0.552	−95	111
Ethylmethyl carbonate (EMC)	2.4	0.65	−55	107
Diethyl carbonate (DEC)	2.82	0.748	−43	126
Dimethyl carbonate (DMC)	3.12	0.585	0.5	90
Propyl acetate (PA)	6.00	0.551	−95	102
Ethyl acetate (EA)	6.02	0.426	−84	77
Methyl acetate (MA)	6.67	0.368	−96	56
Methyl formate (MF)	8.90	0.328	−99	36
γ-Butyrolactone (γ-BL)	41.77	1.727	−43	202
Propylene carbonate (PC)	64.92	2.53	−49	241
Ethylene carbonate (EC, 38 °C)	89.6	1.85	36	248

Source: Reproduced with permission Shiao et al. [14].

is 24.2 mS cm^{-1} at room temperature, it dramatically goes down to 5.6 mS cm^{-1} at −40 °C, only one fifth of the standard value [19]. Thus, it is also an important measure to increase the conductivity of electrolytes at low temperature [20].

On the other hand, researchers ascribed the poor performance of cells at low temperature due to the anodes that are the low value and concentration dependence of the Li^+ diffusivity and limited Li capacity [21, 22]. The diffusion of Li^+ in carbon anodes restricts the performance, while large particles have longer lengths for Li^+ diffusion than small particles, causing a higher concentration polarization on the surfaces of large particles and lower half-cell voltage during the discharge [14, 21].

Electrochemical impedance spectroscopy (EIS) is a powerful technique to research the effects of different parts in cells [23, 24]. In the typical EIS pattern,

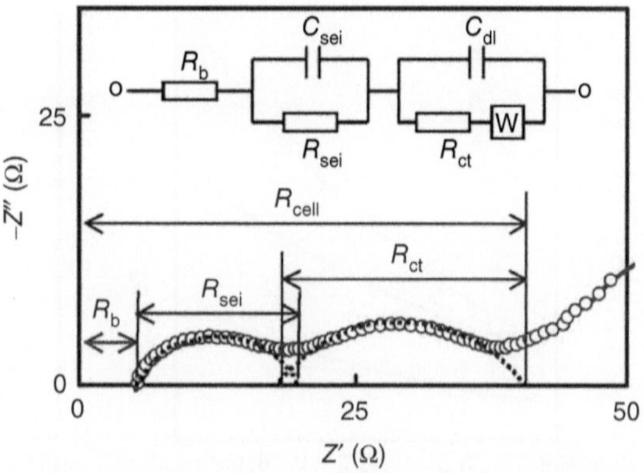

Figure 5.2 Typical EIS pattern and the equivalent circuit used to fit the EIS. Source: Reproduced with permission Zhang et al. [24].

there are usually two partially overlapped semicircles and a straight slopping line at the low frequency end, which can be fitted by an equivalent circuit, as shown in Figure 5.2. R_b is the cell bulk resistance, originated from the electrolyte, separator, and electrodes. R_{sei} and C_{sei} are resistance and capacitance of the solid electrolyte interface (SEI), respectively, on behalf of the first semicircle at high frequencies. R_{ct} and C_{dl} refer to charge-transfer resistance and its relative double-layer capacitance, on behalf of the second semicircle at medium frequencies. W stands for Warburg impedance, which is connected to a combined effect of the diffusion of Li^+ on the electrode–electrolyte interfaces, on behalf of the straight slopping line.

Ratnakumar et al. [25] suggested that the influences of the interfacial properties and characteristics of the SEI on low-temperature performance are more than that of the bulk ion transport in electrolytes. The quality of the SEI plays an important role in the kinetics of lithiation/delithiation and the stability of the interface. Zhang et al. [26] compared discharge curves of cells at various temperatures and found that the operating voltage reduced with the decreasing temperature. The decreased operating voltage was due to the increased electric polarization, which is related to the poor ionic conductivity of the electrolyte and SEI to slow down the cell electrochemical reactions. However, more researches proposed that R_{ct} is the dominant part of the cell resistance. Indeed, R_b and R_{sei} increase with the decrease in temperature, as shown in Figure 5.3a. The increasing trend of R_{ct} is much more obvious than that of R_b and R_{sei} with the decrease in temperature. When the temperature is below −20 °C, R_{ct} accounts for almost 100% of the total cell resistance [26]. Besides, viscosity of the electrolyte will increase and the conduction velocity of ions will slow down, resulting in a mismatch of electron migration velocity in the external circuit. Therefore, the battery will be seriously polarized and the charge–discharge capacity will decrease sharply. Especially, when charging at low temperature, lithium ions easily form lithium dendrites on the anode surface. The uneven deposition of lithium will break the fragile SEI and cause excessive side reactions with the electrolyte, resulting in gas evolution [28].

At low temperature, a typical phenomenon occurs that charging of the discharged cell is more difficult than discharging the charged cell [26]. Figure 5.3b displays the

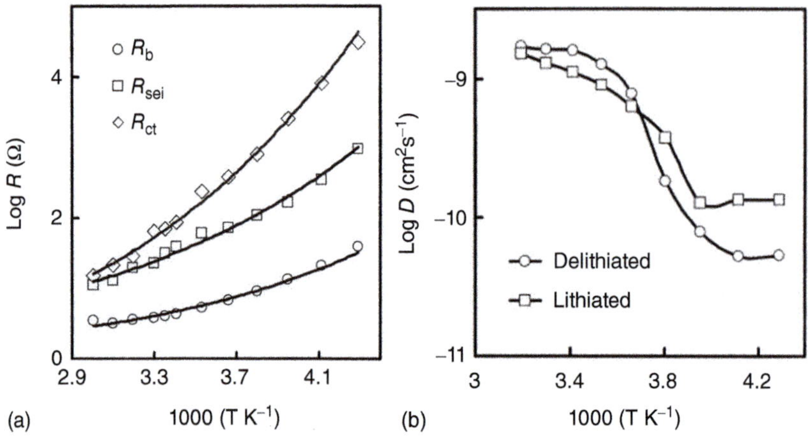

Figure 5.3 (a) Temperature dependence of different resistances. (b) Log D of lithium ions in graphite as a function of 1/T for lithiated and delithiated graphite, respectively. Source: Reproduced with permission Zhang et al. [27].

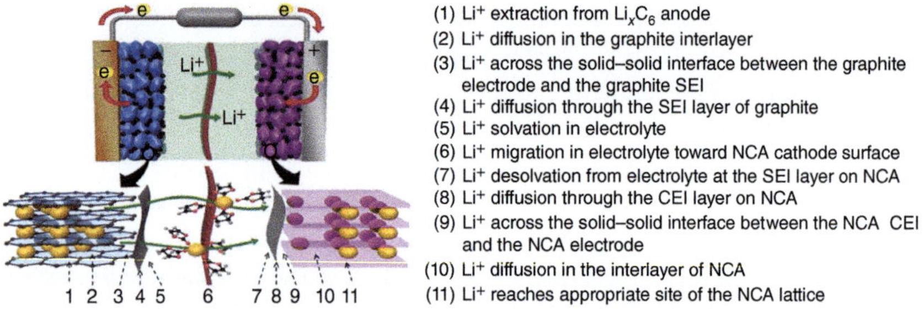

Figure 5.4 Diagram of each step in the discharge process of Li^+ migration. Source: Reproduced with permission Li et al. [32].

effect of temperature on the chemical diffusion coefficient (D) of Li^+ in lithiated and delithiated graphite [27]. The D value sharply decreases below 0 °C, and the D value of delithiated is lower than that of lithiated when the temperature is lower than −20 °C. Electrochemical kinetics of L^+ diffusion on delithiated graphite is slower than that on lithiated one. When the lithium-ion battery is charged immediately after the discharge, due to the high R_{ct} in the fully discharged state, it will cause considerable polarization, and the battery voltage will jump to the preset upper limit in the charging process [29]. This is why the cells are hard to be charged with high rate at low temperature [12, 30]. Faradic impedance reflects the kinetics of cell reactions, which is the combination of R_{ct} and W. The higher the R_{ct}, the slower the kinetics of the Faradic reaction. In short, high R_{ct} is the rate-determining step that dominates the cell performance at low temperature [27, 31].

Furthermore, Xu's group [32] analyzed each step of discharge at low temperature, as shown in Figure 5.4. They confirmed that Li^+ desolvation was the main kinetic obstacle to Li^+ migration at low temperature, rather than Li^+ migration through the

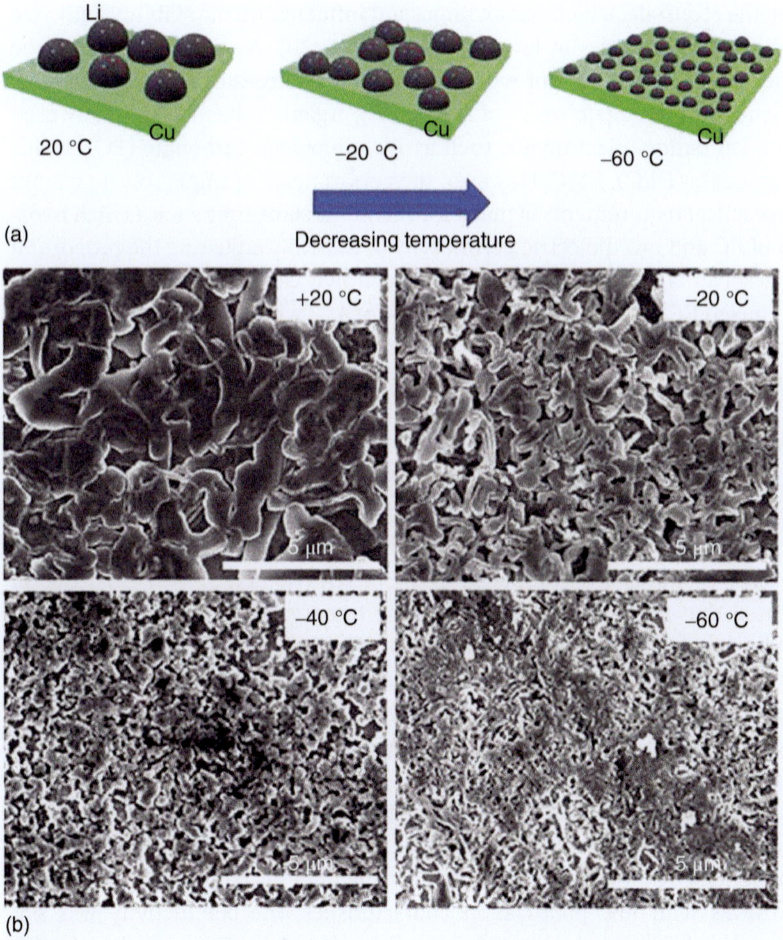

Figure 5.5 (a) Schematic diagram of lithium deposited particles at different temperatures. (b) SEM images of deposited Li on Cu substrates at different temperatures. Source: Thenuwara et al. [36]/with permission of American Chemical Society.

SEI or diffusion in solid electrodes. In addition, the most energy consuming step for Li^+ to enter the electrode is the solvent sheath stripping of Li^+, and the energy barrier has grown more and more obviously when the temperature drops below −20 °C [33–35]. In addition, lithium deposition tends to form smaller particles with decreasing temperature (Figure 5.5). During the plating and striping, the smaller particles, covered by the SEI, are more likely to transform into "dead" Li, resulting in reduction in CE at low temperature [36].

5.1.2 The Improvement of Electrolytes

Electrolyte is an important part of lithium-ion batteries, which not only determines the mobility of ions in electrolyte, but also is related to the stability of the SEI film formed on the electrode, which has an important influence on the stability and cycle life of lithium–ion batteries at low temperature [25, 30]. As the main part of the electrolyte, the choice of solvent will greatly affect the freezing point of the electrolyte. Generally, carbonate series of high-purity organic solvents are widely used in lithium–ion battery electrolytes, such as EC, propylene carbonate (PC), DMC, diethyl carbonate (DEC), EMC. However, the performance of single solvent cannot meet the practical requirements of many aspects at the same time such as high freezing point of EC and low dielectric permittivity of EMC. By adjusting the proportion of different solvents, the freezing point of the electrolyte can be reduced easily. Zhang et al. [27] mixed EC and EMC in a volume ratio of 3 : 7 with 1 M $LiPF_6$ to obtain the electrolyte that did not solidify and could work normally at low temperature. The cells containing this electrolyte could retain about 60% of the capacity at −10 °C. Lim's [37] group reported that capacity retention of the EC–DMC containing 1 M $LiPF_6$ lithium–ion batteries could reach up to 24% at −20 °C. Binary mixed solvents can significantly reduce the freezing point of the electrolyte, but the performance of the cells is poor with low capacity retention rates [38].

To further widen the liquid range and enhance the cell performance, ternary or quaternary mixtures are introduced to the electrolyte [21, 39, 40]. Smart et al. [39] reported that the electrolyte based on the ternary mixture of EC, DEC, and DMC had a better combination in conductivity and surface film characteristics. The interfacial properties of ternary mixtures are more favorable to the intercalation/delamination kinetics of Li^+, especially at low temperature. As shown in Figure 5.6a, after 600 cycles, the cells with 1 M $LiPF_6$ in EC–DMC–DEC (1 : 1 : 1) maintained over 85% capacity retention at −25 °C. Plichta et al. [7] compared EC–DMC–DEC, EC–DMC–EMC, and EC–EMC–DEC in different proportions and found that the electrolytes with EMC had wide liquid ranges at low temperature and high conductivity. Although the electrolyte having high EC content could generate favorable films with low polarization characteristics, the conductivity was not satisfactory. Furthermore, they tested the cells with 1 M $LiPF_6$ in EC–DMC–DEC

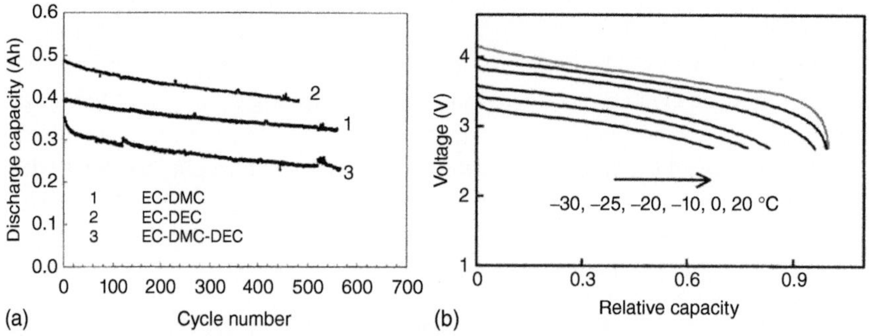

Figure 5.6 (a) Cycle life of lithium–ion batteries with different electrolytes at −25 °C. Source: Reproduced with permission Smart [39]. (b) Voltage–relative capacity curves of Li-ion cells with 1 M $LiPF_6$ 1 : 1 : 3 PC:EC:EMC electrolyte at various temperatures. Source: Reproduced with permission Zhang et al. [41].

(1 : 1 : 1) at −40 °C and obtained 52% capacity retention [42]. PC is usually added to inhibit the crystallization of EC at low temperature, because PC has not only good intersolubility with EC but also low freezing point. The main reason limiting the use of PC is that PC is easy to be co-embedded with lithium ions into graphite anode, which peels off the graphite layer and leads to a decline in the battery cycle performance. Zhang et al. [41] regulated the different ratios of PC, EC, and EMC to eliminate the defects of single solvent and found that 1 M $LiPF_6$ 1 : 1 : 3 PC:EC:EMC electrolyte greatly improved the low-temperature performance of the cells by the higher conductivity SEI film on the interfaces, as shown in Figure 5.6b. Mandal et al. [43] reported that EC–DMC–EMC in a weight ratio of 15 : 37 : 48 with 0.9 M LiTFSI had high conductivity about 2.0 mS cm^{-1} at −40 °C. In development, quaternary solvent systems combine more advantages to further enhance the properties of electrolytes [26, 44]. Huang et al. [21] reported that the electrolyte with EC:DEC:DMC:EMC (3 : 5 : 4 : 1)/0.8 M $LiPF_6$ was 100% liquid even at 42.5 °C, while other electrolytes with different ratios became solid. Approximately 87% capacity was maintained at −40 °C. Smart et al. [45] reported that the most promising electrolyte formulations of EC:DEC:DMC:EMC were 1 : 1 : 1 : 2 and 1 : 1 : 1 : 3. The low EC-content electrolytes exhibited higher limiting current densities at low temperature. The cells with 1 M $LiPF_6$ in EC–DEC–DMC–EMC (1 : 1 : 1 : 3) delivered 74.6% of room temperature capacity at −40 °C, 68.9% at −50 °C, and 35% even at −70 °C.

Aliphatic esters have ultralow freezing points and can be introduced as cosolvents into electrolytes [18, 37]. For example, methyl formate (MF) was mixed with EC by Ein-Eli et al. [19] because of its freezing point of −99 °C and high stability upon oxidation. The electrolytes with 1 M $LiAsF_6$ in EC–MF (1 : 3) displayed a conductivity of 8.4 mS cm^{-1} at −40 °C, higher than 5.6 mS cm^{-1} in EC–MF (1 : 1). Sazhin et al. [37] reported that the cells with the electrolyte containing 1 M $LiPF_6$ in EC–DMC–MA (30 : 30 : 40) maintained 94% discharge capacity at −20 °C. However, both MF and MA are "aggressive" solvents, which will erode the anode if there is no other protection. For that reason, Shiao et al. [14] introduced toluene into the electrolyte that contained MF or MA to enhance the cycle performance of batteries. By optimizing the composition of these mixed solvents and changing the salt concentration, they achieved the preservation of rates and capacities at low temperature [14]. The electrolytes containing 1 M $LiPF_6$ in EC–DMC–MA–toluene (1 : 1 : 1 : 1) exhibited a conductivity of 1.86 mS cm^{-1} and cycled ~160 times steadily at −40 °C. Smart et al. [46] examined a series of ester cosolvents in the formulation 1 M $LiPF_6$ in EC–EMC–x (20 : 60 : 20), where x = methyl propionate (MP), ethyl propionate (EP), methyl butyrate (MB), ethyl butyrate (EB), propyl butyrate (PB), and butyl butyrate (BB). As shown in Figure 5.7, the best performance delivered 63% of room temperature capacity at −40 °C from the electrolyte with MP. Other cosolvents also yielded improvement to a certain extent compared to carbonate-based electrolytes.

As another important component of the electrolyte, lithium salts directly affect the conductivity of Li^+ and the stability of SEI film. In order to ensure good low-temperature performance, lithium salts that are easily soluble in organic solvents and have high chemical stability and thermal stability are selected. $LiPF_6$ is the

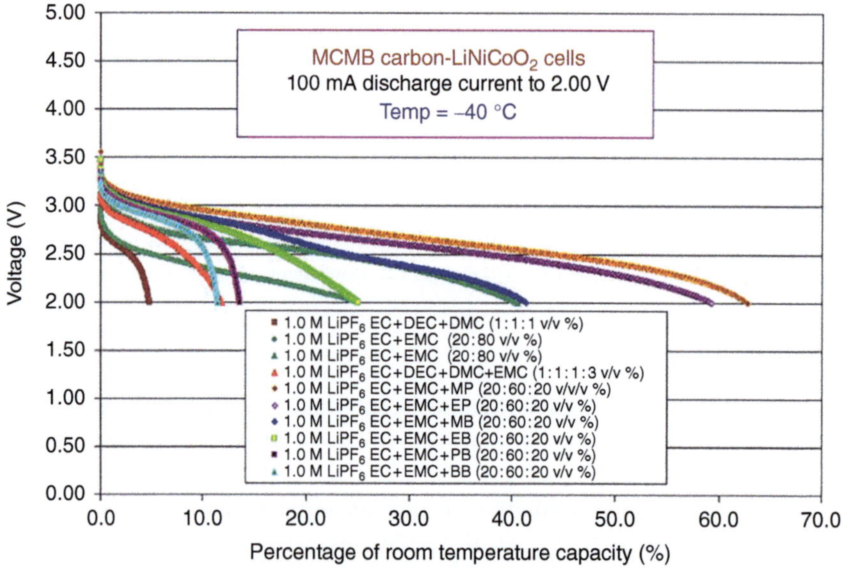

Figure 5.7 Discharge capacity of cells containing different electrolytes at −40 °C. Source: Reproduced with permission Smart et al. [46].

most commercially used salt due to its high conductivity and low cost, but it easily hydrolyzes to generate HF, which will destroy the electrode active material and the corrode collector, resulting in capacity attenuation. In addition, $LiPF_6$ usually needs to be dissolved in an electrolyte containing EC to form a stable SEI film, which limits its application at low temperature due to high freezing point of EC. Many lithium salts were explored to replace $LiPF_6$. Zhang et al. [29] used lithium tetrafluoroborate ($LiBF_4$) to confect the electrolyte with 1 M $LiBF_4$ in PC–EC–EMC (1 : 1 : 3). The cells with this electrolyte maintained 86% capacity retention at −30 °C compared to 72% that used $LiPF_6$ (Figure 5.8a). Analysis of cell impedance implied that R_{ct}

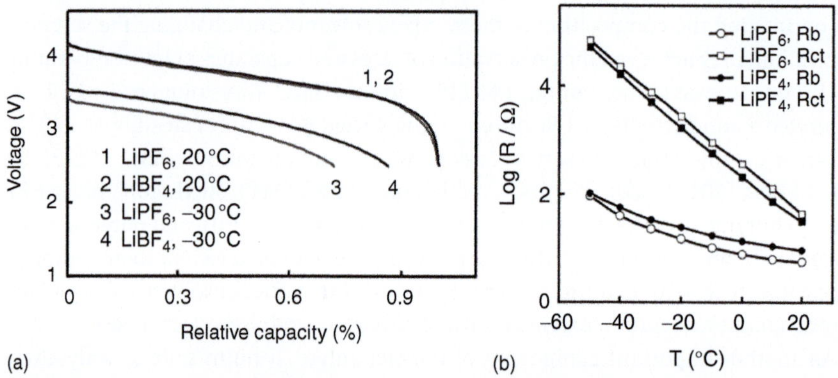

Figure 5.8 (a) Discharge voltage and capacity of cells with different salts. (b) Temperature dependence of R_b and R_{ct} of cells with different salts. Source: Reproduced with permission Zhang et al. [29].

is much more sensitive to temperature than R_b, as shown in Figure 5.8b, and the improvement by $LiBF_4$ originated from the reduced R_{ct}. Lithium bis(oxalato)borate (LiBOB) is cheaper than any other Li salts and regarded as a potential replacement for $LiPF_6$ [47]. Using PC as the solvent instead of EC to form LiBOB electrolyte not only can stabilize graphite anode but also have unique properties that $LiPF_6$ does not have, and can broaden the application range of lithium-ion batteries at low temperature. Lithium oxalyldifluoroborate (LiODFB) has the common advantages of $LiBF_4$ and LiBOB, because its chemical structure is composed of half of $LiBF_4$ and LiBOB molecular structures [48]. Compared with LiBOB, LiODFB is more soluble in straight-chain carbonate solvents, and lithium-ion batteries using LiODFB as electrolyte display better low-temperature performance and high rate discharge performance [49]. Fu et al. reported that the electrolyte with 1 M LiODFB in EC–PC–EMC (1 : 1 : 3) generated a denser and more compacter surface film on the cathode. Lithium bis(trifluoromethylsulfonyl)imide (LiTFSI) is stable, safe, and nontoxic as well as has high conductivity in any medium, so it is the ideal candidate to replace $LiPF_6$ [50]. Mandal et al. [43] reported that the charge-transfer resistance of an electrolyte with LiTFSI is almost an order of magnitude lower than that with $LiPF_6$. Xia's group [51] reported a simple electrolyte with 2 M LiTFSI in EA, which decreased the freezing point to the ultralow temperature of −91 °C and exhibited a sufficient ionic conductivity of 0.2 mS cm^{-1} at −70 °C. However, the electrolyte still could not overcome the problems such as the sluggish desolvation of Li^+ and the absence of compact SEI. By further studies, they optimized the electrolyte containing 5 M LiTFSI in EA-DCM (1 : 4) to obtain local high concentrations [52]. The optimized electrolyte had higher ionic conductivity up to 0.6 mS cm^{-1} and further enhanced the performance of cells at −70 °C.

Similar to mix solvents, the mixture of lithium salts to form blended-salt electrolytes can deliver synergistic effects that are beneficial to improve the properties of electrolytes. Zhou et al. [53] combined $LiBF_4$ and LiODFB in EC–DMC–EMC (1 : 1 : 1) and explored the effects of different ratios of the two salts. As shown in Figure 5.9a, the temperature decrement produced a decrease in ionic conductivity. The electrolytes with $LiBF_4$ displayed better conductivity at low temperature because BF_4^- has a smaller Van der Waals volume than $ODFB^-$. When the ratio was 4 : 1, the cells delivered the highest capacity retention after 100 cycles at −20 °C. In addition, Zhao et al. [54] reported an electrolyte system of LiODFB/$LiBF_4$ (1 : 1) in EC–DEC–DMS (1 : 2 : 1) as an optimized electrolyte that exhibited 55.7% capacity retention at −40 °C (Figure 5.9b). Cui's group [55] formulated 0.6 M LiTFSI–0.4 M LiTFPFB–0.05 M $LiPO_2F_2$ in PC–EC–EMC (1 : 1 : 3) as electrolyte. The addition of $LiPO_2F_2$ contributed to generate a compact SEI layer composed of inorganic lithium species and P–O species, which were the key elements in inhibiting the growth of Li dendrites. Thenuwara et al. [56] reported on a dual-salt electrolyte that mixed 0.8 M LiTFSI and 0.2 M $LiNO_3$ and obtained high CE at −60 °C. Xia's group [57] blended 0.8 M LiTFSI and 0.2 M LiODFB and regulated with adiponitrile to formulate the electrolyte that enhanced the cell performance.

The use of additives is one of the most convenient and efficient strategies to improve the properties of electrolytes [58]. Normally, the amount of additives in

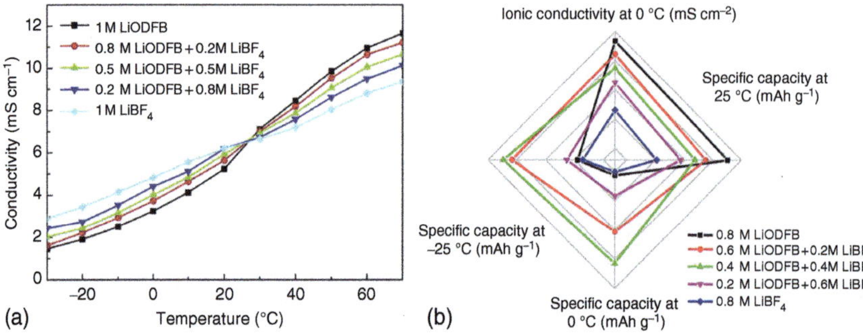

Figure 5.9 (a) Conductivity of different electrolyte systems at different temperatures. Source: Reproduced with permission Hongming et al. [53]. (b) A "radar" plot that compares different electrolyte systems. Source: Reproduced with permission Qiuping et al. [54].

electrolytes is very low but yields significant improvement. Selecting additives with specific functions can enhance the cell performance at low temperature. Herreyre et al. [59] used 5% vinylene carbonate (VC) as additive to improve stability of the passivating layer and enhanced the cell performance at −40 °C. Smart et al. [60] reported that the addition of 4% fluoroethylene carbonate (FEC), 2% lithium oxalate, 2% VC, and 0.1 M LiBOB in 1.2 M $LiPF_6$ in EC–EMC–MB (1 : 1 : 3) all greatly increased the capacity retention of cells at −40 °C. Among that, 2% VC and 0.1 M LiBOB exhibited the best performance of 53% capacity retention, as shown in Figure 5.10. The results implied that L^+ intercalation and deintercalation kinetics were more favorable with the cells containing the additives. Wagner et al. [61] demonstrated methyl vinyl sulfone (MVS) and ethyl vinyl sulfone (EVS) as film-forming additives, which could simultaneously generate a stable SEI on the anode and exert a slight impact on cathode performance. Yang et al. [62] introduced 1% $LiPO_2F_2$ into the electrolyte and obtained a conductive and stable SEI, which was rich in LiF and $LiCO_3$ components to decrease impedance, resulting in 71.9% capacity retention at −20 °C and 57.93% capacity retention at −30 °C.

Apart from conventional electrolyte systems, there are some unique strategies to improve the performance of cells at low temperature [2, 63]. Smart et al. [64] using a gel polymer system obtained about 99% capacity retention with a C/10 discharge rate at −40 °C. They also found that higher molecular weight esters tend to display improved stability electrochemically and are preferred over lower molecular weight esters (such as methyl acetate and ethyl acetate) for robust low-temperature performance over the cell life, although the latter cosolvents can often yield excellent low-temperature capability in the initial stages of the cell life. Gao et al. [65] reported an interface manipulation approach to stabilize Li metal anodes at low temperature by using an electrochemically active monolayer (EAM) to regulate the interfacial chemical environment. As a result, a multilayered SEI was formed that contained a LiF-rich inner phase and an amorphous outer layer (Figure 5.11a), which rendered stable Li deposition at −60 °C. Electrolytes are primarily considered liquid

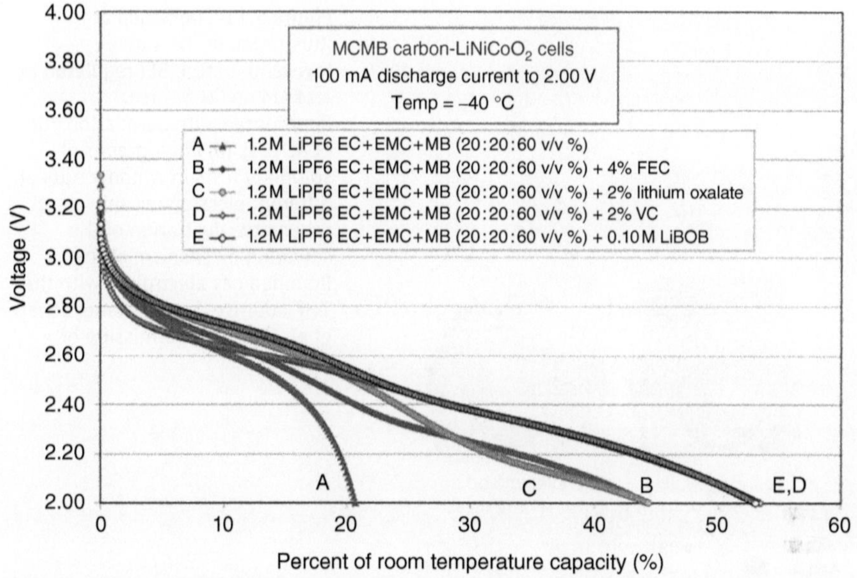

Figure 5.10 Capacity retention of cells containing different additives in the electrolyte at −40 °C. Source: Reproduced with permission Smart Marshall et al. [60].

or solid, allowing ion diffusion to occur through coordinated interactions with the surrounding dielectric medium. It is generally believed that the intermolecular interaction between gas molecules at room temperature is poor, resulting in lack of consolidated phase. However, Meng's group creatively employed gaseous electrolytes, by using pressurized hermetically sealed cell setups, to vary the effects of polarity and ion coordination within electrolyte solvents. They found that monofluorinated solvents such as fluoromethane (FM) have the ability to solvate lithium salts, and it is proved that the formation of highly fluorinated SEI can stabilize the deposition of lithium metal [67]. Furthermore, they selected tetrahydrofuran (THF) to tailor the electrolyte, which could coordinate with Li^+ to enhance solvation and transport, increasing the coulombic efficiency (CE) of Li deposition from 97% to 99.6% [66]. Besides, LiTFSI could fully coordinate with THF (Figure 5.11b) to increase the solubility with the addition of THF in a 1 : 1 salt-to-additive ratio, while only 0.1 M in previous electrolyte. This work was further followed up by their group by using an analogous approach with FM liquefied gas electrolyte and acetonitrile cosolvent [68]. The optimized electrolyte had high conductivity of 4.8 mS cm^{-1} at −78 °C and rendered high CE of 99.4% at high current density of 3 mA cm^{-2}.

5.2 High-Temperature Electrolytes

In the use of battery conditions, high temperature often encounters a situation. Some industrial applications such as geosatellite positioning of freight containers and military applications need batteries that properly operate at high temperature

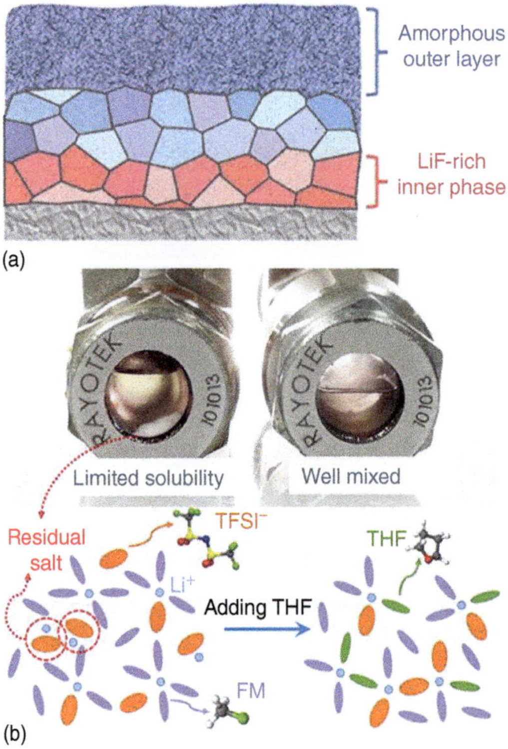

Figure 5.11 (a) Schematic illustration of the stable low-temperature SEI regulated by an EAM on Cu. Source: Reproduced with permission Yue et al. [65]. (b) Photographs of solubility tests in window cells of different electrolytes and schematic illustration of the solvation mechanism of the liquefied gas electrolyte with the THF additive. Source: Yangyuchen et al. [66]/with permission of Elsevier.

[3, 69]. The optimal operating temperature for conventional batteries ranges from 15 to 35 °C, as shown in Figure 5.12 [70]. Different from slow kinetics at low temperature, the main problems at high temperature are acceleration of side reactions and aging of batteries [71]. The capacity of batteries cycled at high temperature fades more severely than batteries cycled at room temperature and a more limited capacity range [72]. Therefore, improving battery stability is the key point to study at high temperature.

5.2.1 The Limitations of Battery Performance at High Temperature

At high temperature, the stability of the electrolyte and the thermal stability between the electrolyte and anode and cathode interfacial films are the main factors affecting high-temperature electrolytes [72, 73]. Decomposition of electrolytes and damage to the interfacial film at high temperature will lead to poor performance of the battery at high temperature. At high temperature, each component of the electrolyte and the trace water in the electrolyte may participate in the interfacial chemical reaction. High temperature also aggravates the interfacial chemical reaction, resulting in poor stability of the interfacial film [74, 75].

Aging is also a serious problem during the application of batteries at high temperature, which not only affects the performance but also reduces the lifetime. Increasing the operating temperature of cells above the optimal scope will accelerate the aging

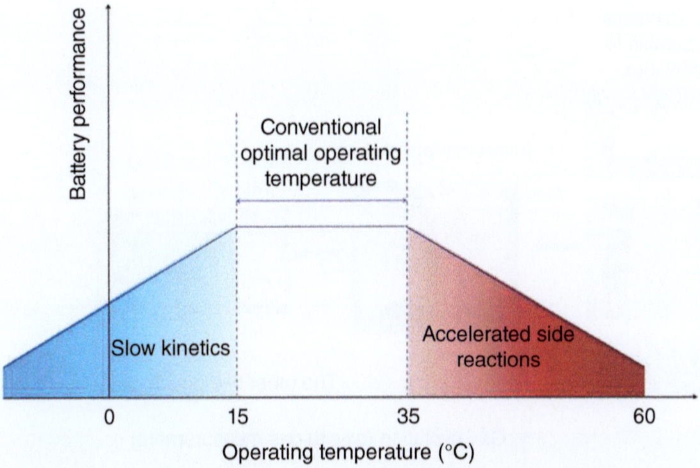

Figure 5.12 The cell performance on different operating temperature windows. Source: Reproduced with permission Jiangyan et al. [70].

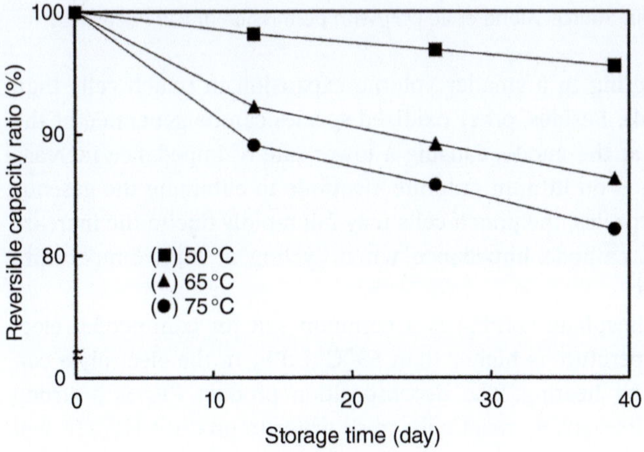

Figure 5.13 Reversible capacity with various storage times at different temperatures. Source: Reproduced with permission Araki et al. [78].

process and lead to degradation [76, 77]. As shown in Figure 5.13, the reversible capacity of cells decays obviously while stored at high temperature. After 40 days stored at 75 °C, the reversible capacity loss was about 20% [78]. The SEI on the stored anode was not uniform and formed by the decomposed products from $LiPF_6$, polymerized EC, and Mn dissociated from the cathode material, with the thickness varying from 40 to 200 nm [78].

Being thermally activated, the self-discharge rate increases with storage temperature. Meanwhile, electrolyte oxidation becomes increasingly severe, leading to gaseous products and other oxidized species. Xiong et al. [79] found that a large amount of CO_2 was generated at the cathode side, while CO_2 would be reduced

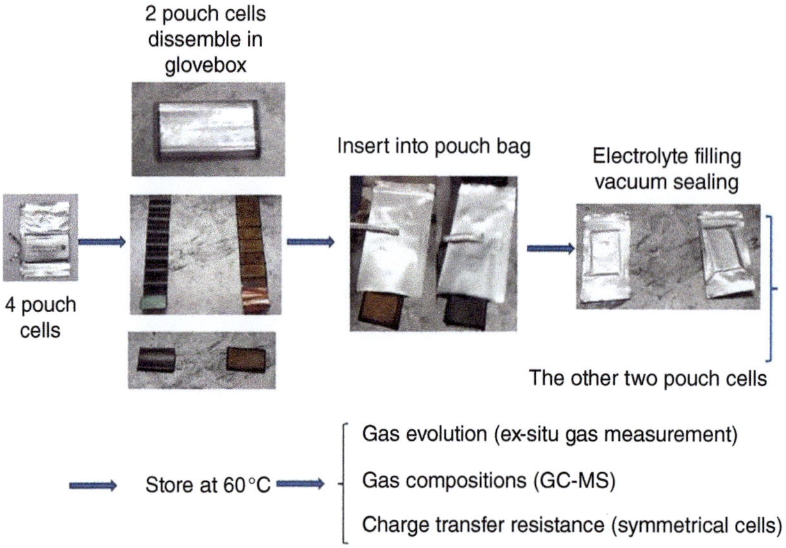

Figure 5.14 Photographs of steps and a flow path of the process used to make and test pouch cells and pouch bags. Source: Xiong et al. [79]/with permission of IOP Publishing.

at the anode side, resulting in a smaller volume expansion in pouch cells than pouch bags (Figure 5.14). Besides, other oxidized species can be generated at the cathode and expended at the anode, causing a lower rate of impedance increase at the cathode. If there is no lithium graphite electrode to eliminate the gaseous products and oxidized species, the pouch cells may fail rapidly due to the increase in gas production and cathode impedance when cycling at high temperature [79].

Lithium hexafluorophosphate ($LiPF_6$) is a common salt for commercial electrolytes. When the temperature is higher than 55 °C, $LiPF_6$ in the electrolyte can easily be decomposed by heating. The decomposition product PF_5 is a strong Lewis acid, which can irreversibly react with trace water to produce HF. HF will corrode the cathode material and cause severe loss in battery capacity. Therefore, the selection of appropriate lithium salt is also an important step to improve the cell performance of cells at high temperature. All these factors lead to a reduction in high-temperature cycle stability of the battery. Moreover, the structure of some cathode materials is unstable at high temperature, resulting in dissolution of transition metals [80, 81]. During cycling, Li^+ will intercalate into the cathode lattices, causing the phase change of cathodes, which reduces the available capacity [82]. In addition, the migration of PVdF binder also affects the cell performance at high temperature. As shown in Figure 5.15, binder migration caused the generation of a PVdF layer at the extreme surface of the cathode and lower Li reintercalation in part 1 [83].

At high temperature, according to the Nernst equation, the anode and cathode potentials shift negatively, so the electronic energy levels of the electrodes also jump. Meanwhile, the highest occupied molecular orbital (HOMO) and the lowest unoccupied molecular orbital (LUMO) also change with the temperature increase.

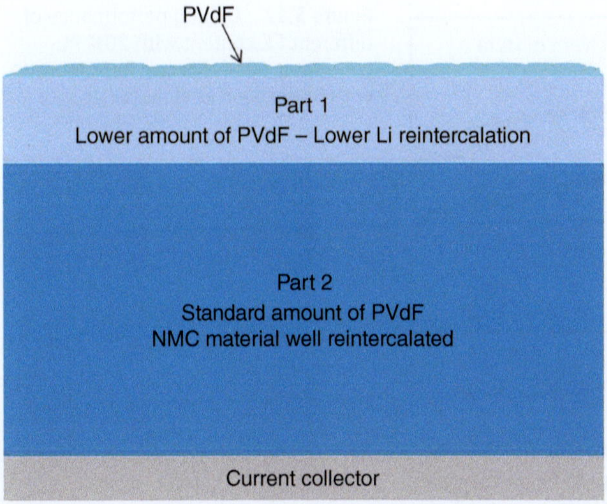

Figure 5.15 Schematic diagram of the cathode after cycling at 120 °C. Source: Reproduced with permission Lucille et al. [83].

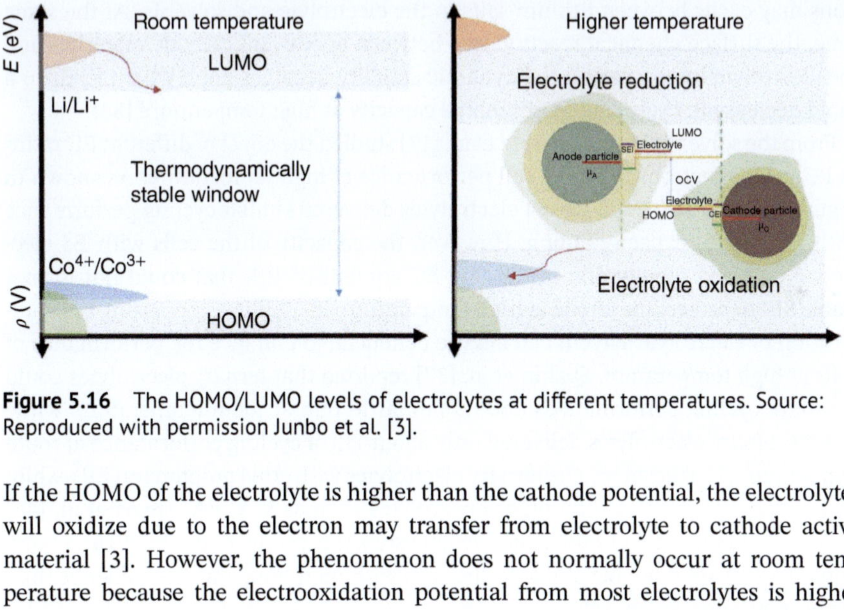

Figure 5.16 The HOMO/LUMO levels of electrolytes at different temperatures. Source: Reproduced with permission Junbo et al. [3].

If the HOMO of the electrolyte is higher than the cathode potential, the electrolytes will oxidize due to the electron may transfer from electrolyte to cathode active material [3]. However, the phenomenon does not normally occur at room temperature because the electrooxidation potential from most electrolytes is higher than 5 V. From the thermodynamic point of view, the electronic energy levels of the anode and cathode increase and the gap between HOMO/LUMO narrows at high temperature. Thus, the original SEI and CEI are unstable to support at high temperature, as shown in Figure 5.16 [3]. Besides, the superfluous side reactions are not conducive to the formation of the SEI and CEI, but will have a negative impact on the cell performance. Kinetically, the temperature increases with 10 °C, the reaction rate doubles, if the activation energy of the reaction follows the Arrhenius relation.

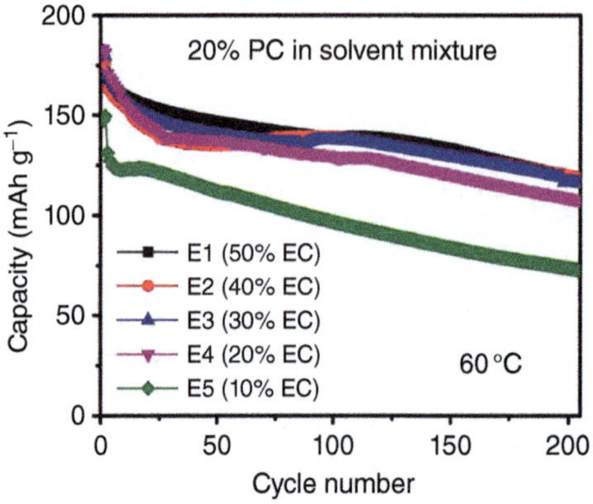

Figure 5.17 Cycling performance of different EC content with 20% PC content at 60 °C. Source: Reproduced with permission Li et al. [17].

5.2.2 The Improvement of Electrolytes

The main problems of an electrolyte at high temperature are the chemical decomposition of the electrolyte itself and the loss of surface chemical passivation mechanism between the electrolyte and anode and cathode. At high temperature, chemical reactions may occur between lithium salts in the electrolyte and solvents. At the same time, the surface chemical reaction rate between anode and cathode materials and the electrolyte increases and the dynamic stability becomes poor, which leads to a rapid decrease in cycle charge–discharge capacity at high temperature [84].

From the solvent point of view, Li et al. [17] studied the effect of different EC ratios in PC-based electrolytes on the cell performance at high temperature. As shown in Figure 5.17, the cells with E1–E4 electrolytes delivered similar cycling performance with only slight capacity fading. However, the capacity of the cells with E5 electrolyte declined rapidly due to the low EC content of 10% that could not form a stable SEI to protect the anode at high temperature. In addition, increasing the composition of electrolyte solvent can also be beneficial to enhance the performance of cells at high temperature. Sazhin et al. [37] reported that ternary electrolytes could maintain cycling performance at 50 °C similar to that at room temperature, while at 50 °C binary electrolytes delivered only about 1/3 of cycling performance at room temperature. Furthermore, the ternary electrolytes with ethyl propionate (EP) exhibited better performance than other electrolytes such as MA, EA. Smart et al. [46] also reported that higher molecular weight esters (i.e. propyl butyrate, PB and butyl butyrate, BB) are more beneficial to the cells with high-temperature resilience than lower molecular weight esters.

Due to the drawbacks such as poor thermal stability of $LiPF_6$, researchers have done a number of works to find the replacement for $LiPF_6$ [55, 57, 85]. For example, LiBOB has good thermal stability and excellent ability to passivate Al and generate a stable film layer on the surface of both electrodes. $LiBF_4$ can enhance the cell performance by lowering down the charge-transfer resistance. Inspired by this,

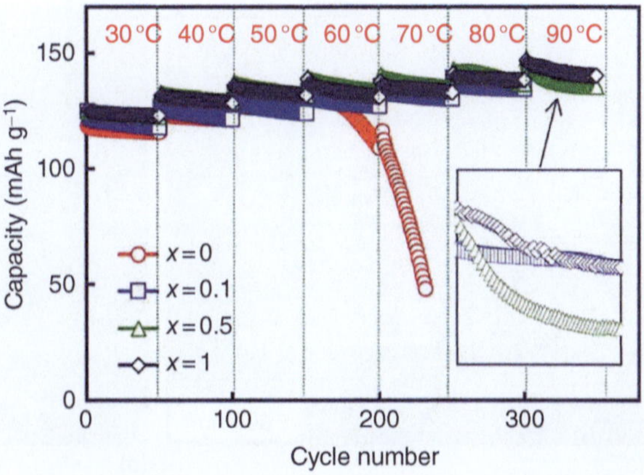

Figure 5.18 Cycling performance of 1 M (1−*x*)$LiBF_{4-x}$LiBOB electrolytes at different temperatures. Source: Reproduced with permission Zhang et al. [4].

Zhang et al. [4] mixed LiBOB and $LiBF_4$ with different ratios in EC–PC–EMC (1 : 1 : 3) as high-temperature electrolytes to cycling test for Li/$LiFePO_4$ cells. The results showed a stable cycling performance of LiBOB without visible fading from 30 to 80 °C (Figure 5.18). Even up to 90 °C, the capacity fading of LiBOB was slight and the cells still could be cycled. The authors attributed the excellent performance to the outstanding thermal stability of LiBOB salts and the inhibitory effect on the dissolution of Fe(II) ions from $LiFePO_4$ [4]. What's more, the addition of just a small amount of LiBOB could remarkably enhance the cell performance at high temperature, while pure $LiBF_4$ could not stably cycled above 40 °C.

A novel lithium salt, LiODFB, has drawn wide attention due to its outstanding advantages such as thermal stability and improvement of SEI. Li et al. [86] reported an electrolyte prepared by dissolving 1 M LiODFB in EC–PC–DMC (1 : 1 : 3). Compared to $LiPF_6$-based electrolytes, this LiODFB-based electrolyte had a relatively higher conductivity of 8.25 mS cm^{-1} at room temperature and a better cycling performance at 65 °C with 88% capacity retention after 100 cycles. Fu et al. [48] also reported an electrolyte containing 1 M LiODFB in EC–PC–EMC (1 : 1 : 3) and assembled the cells with $LiMn_2O_4$ cathode for test. From the SEM images of $LiMn_2O_4$ cathodes cycled at 60 °C, the cathode surface was covered by an uneven and unconsolidated film when cycled in $LiPF_6$-based electrolytes (Figure 5.19a). In sharp contrast, a uniform and compacted film was observed on the cathode surface that cycled in LiODFB-based electrolytes (Figure 5.19b). What's more, LiODFB-based electrolytes could also inhibit the dissolution of Mn ions because $ODFB^-$ anions could capture dissolved Mn ions and form a denser surface film on the cathode surface to prevent Mn ions from continuing to dissolve, especially at high temperature. The ICP result in Figure 5.19c shows that only less than 0.1 ppm of Mn^{2+} was detected in LiODFB-based electrolytes, while almost 13 ppm of Mn^{2+} was observed in $LiPF_6$-based electrolytes [48]. Han et al. [87] introduced lithium (fluorosulfonyl)(nonafluorobutanesulfonyl)imide (LiFNFSI) as a novel salt to enhance the high-temperature resilience of cells. LiFNFSI displayed a superior

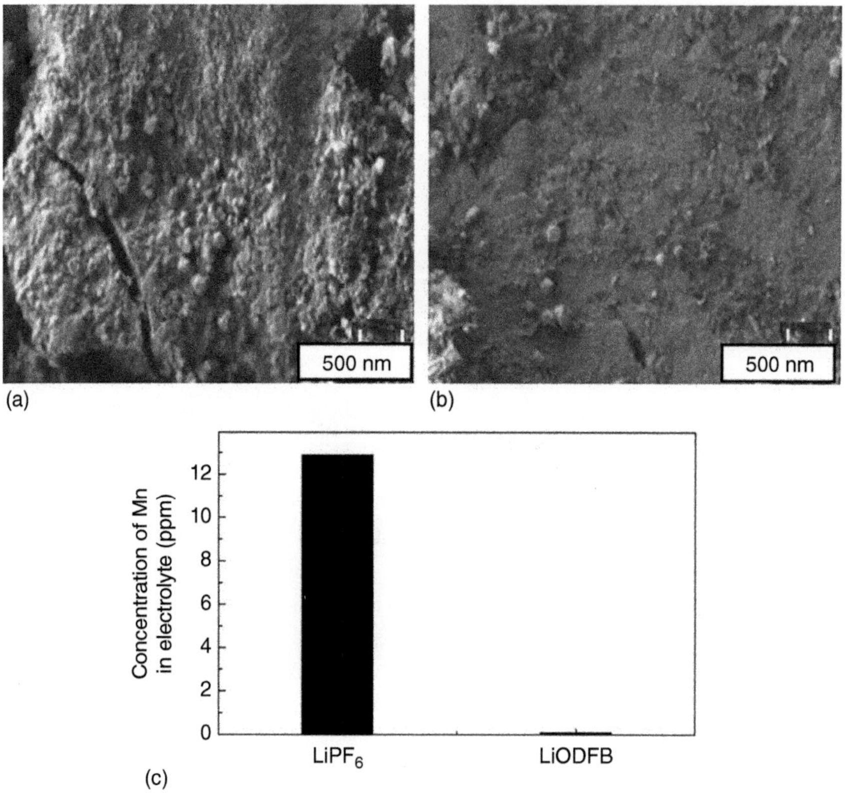

Figure 5.19 SEM images of the $LiMn_2O_4$ electrode after cycling in $LiPF_6$-based (a) and LiODFB-based electrolyte (b) at 60 °C. The concentration of Mn^{2+} in electrolytes with different salts (c). Source: Fu et al. [48]/with permission of Elsevier.

thermal stability (220 °C) over $LiPF_6$, conductivity close to $LiClO_4$, adequate electrochemical stability, and no corrosion of Al.

The introduction of additives is also a feasible and convenient strategy to improve high-temperature performance of cells. Smart et al. [60] investigated many additives and the results of the cycling performance at high temperature are displayed in Figure 5.20. The addition of VC and lithium oxalate enhanced the highest capacity retention at 60 and 80 °C. VC could provide favorable film characteristics at the cathode and reduce film and charge-transfer resistances. Markovsky et al. [88] used a Li-organoborate complex (Li-OBC) as additives, which reduced the capacity loss from 14% (additive-free) to 3% at 60 °C. Nam et al. [89] introduced diphenyloctyl phosphate (DPOF) and tris(2,2,2-trifluoroethyl) phosphite (TTFP) as flame-retardant additives into an electrolyte composed of 1.15 M $LiPF_6$ in EC–EMC (3 : 7). The results showed that the cells with DPOF delivered superior discharge capacity and capacity retention both at room temperature and elevated temperature (40 °C) by reducing the internal impedance. However, TTFP had a negative impact on the original electrolyte. Chang et al. [90] reported that the addition of 0.1 M trimethyl borate (TMB) could improve the cycling performance

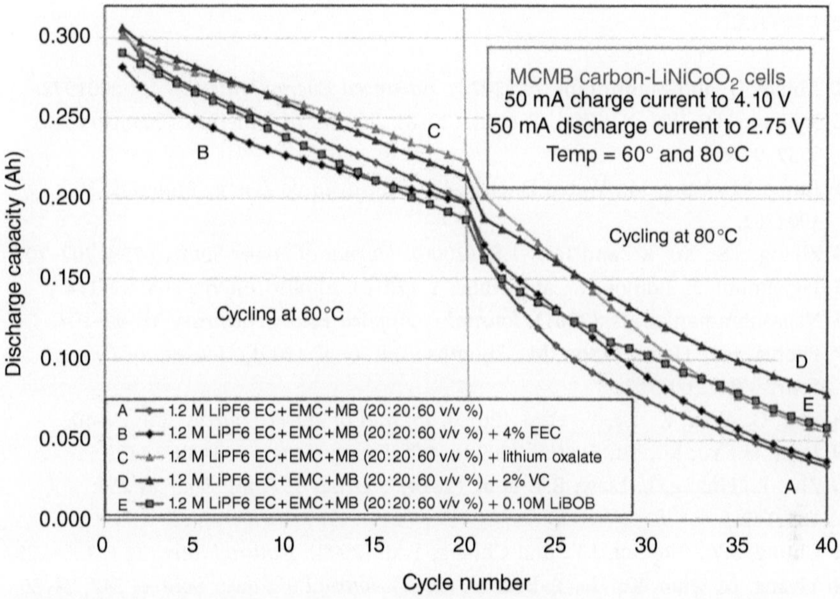

Figure 5.20 Cycling performance of cells with and without additives at 60 and 80 °C. Source: Reproduced with permission Smart Marshall et al. [60].

of cells with 53% capacity retention after 100 cycles at 65 °C. TMB could suppress thermal decomposition of $LiPF_6$-based electrolytes at high temperature and reduce the reaction of impurities with electrolytes.

5.3 Prospective

With the deepening of the battery market, there will be more and more conditions of battery to operate under extreme temperatures. The limitations at low temperature focus on the sluggish kinetics including low diffusivity, high barrier for Li^+ desolvation and Li migration in SEI. At high temperature, chemical decomposition of electrolytes and the loss of surface chemical passivation mechanism are the main problems. At present, a number of researches have been reported to overcome the obstacles and enhance the performance of cells at high or low temperature separately. However, in practical applications such as deep space exploration or military equipment, batteries may need to be operated alternatively at high and low temperature. Therefore, development of electrolytes that can work at both high and low temperature simultaneously is a promising direction for research at extreme temperature. To widen the temperature range of electrolytes that can adapt at both high and low temperature, the electrolyte optimization strategies may be based on the following: (i) choosing lithium salts that have preferable thermal stability and high conductivity; (ii) selecting the solvents with low freezing point and high flash point or mix some specific cosolvents; and (iii) introducing multifunctional additives to stabilize SEI, help desolvation, and reduce interface impedance.

References

1 Abhay, G. and Manthiram, A. (2020). *Advanced Energy Materials* 10: 2001972.
2 Ruihuan, Q., Wei, Y., Zhai, T. et al. (2018). *Journal of Materials Chemistry A* 6: 9737–9746.
3 Junbo, H., Yang, M., Wang, D. et al. (2020). *Advanced Energy Materials* 10: 1904152.
4 Zhang, S.S., Xu, K., and Jow, T.R. (2006). *Journal of Power Sources* 159: 702–707.
5 Jaguemont, J., Boulon, L., and Dubé, Y. (2016). *Applied Energy* 164: 99–114.
6 Nagasubramanian, G. (2001). *Journal of Applied Electrochemistry* 31: 99–104.
7 Plichta, E.J., Hendrickson, M., Thompson, R. et al. (2001). *Journal of Power Sources* 94: 160–162.
8 Zhu, G., Wen, K., Lv, W. et al. (2015). *Journal of Power Sources* 300: 29–40.
9 John, H., Yu, M., Yu, S. et al. (2020). *ACS Energy Letters* 5: 1438–1447.
10 Zhe, L., Huang, J., Liaw, B.Y. et al. (2014). *Journal of Power Sources* 254: 168–182.
11 Chung, S.Y., Bloking, J.T., and Chiang, Y.M. (2002). *Nature Materials* 1: 123–128.
12 Yiyang, L., Qian, K., He, Y.-B. et al. (2017). *Journal of Power Sources* 342: 24–30.
13 Qian, H., Yang, Z., and Mao, J. (2017). *Ionics* 23: 803–811.
14 Shiao, H.C., Chua, D., Lin, H.P. et al. (2000). *Journal of Power Sources* 87: 167–173.
15 Smart, M.C., Whitacre, J.F., Ratnakumar, B.V. et al. (2007). *Journal of Power Sources* 168: 501–508.
16 Xu, K., Lam, Y., Zhang, S.S. et al. (2007). *Journal of Physical Chemistry C* 111: 7411–7421.
17 Li, Q., Jiao, S., Luo, L. et al. (2017). *ACS Applied Materials and Interfaces* 9: 18826–18835.
18 Smart, M.C., Ratnakumar, B.V., and Surampudi, S. (2002). *Journal of the Electrochemical Society* 149: A361.
19 Eineli, Y., Chadha, R., Tj, B. et al. (1997). *Journal of the Electrochemical Society.*
20 Fadhel, A. and Fedkiw, P.S. (2010). *Journal of Power Sources* 195: 7627–7633.
21 Huang, C.K. and Sakamoto, J.S. (2000). *Journal of the Electrochemical Society* 147: 2893.
22 Nobili, F., Mancini, M., Dsoke, S. et al. (2010). *Journal of Power Sources* 195: 7090–7097.
23 Zhang, S.S., Xu, K., and Jow, T.R. (2006). *Electrochimica Acta* 51: 1636–1640.
24 Zhang, S.S., Xu, K., and Jow, T.R. (2004). *Electrochimica Acta* 49: 1057–1061.
25 Rtnakumar, B.V., Smart, M.C., and Surampudi, S. (2001). *Journal of Power Sources* 97: 137–139.
26 Zhang, S.S., Xu, K., and Jow, T.R. (2003). *Journal of Power Sources* 115: 137–140.
27 Zhang, S.S., Xu, K., and Jow, T.R. (2002). *Electrochimica Acta* 48: 241–246.
28 Liu, Q.Q., Xiong, D.J., Petibon, R. et al. (2016). *Journal of the Electrochemical Society* 163: A3010–A3015.
29 Zhang, S.S., Xu, K., and Jow, T.R. (2002). *Electrochemistry Communications* 4: 928–932.

30 Wonyoung, C., Kim, S.-J., Park, I.-T. et al. (2013). *Journal of Alloys and Compounds* 563: 249–253.

31 Sung-Jun, Y., Myung, S.-T., and Sun, Y.-K. (2014). *Journal of the Electrochemical Society* 161: A1514–A1520.

32 Li, Q., Lu, D., Zheng, J. et al. (2017). *ACS Applied Materials and Interfaces* 9: 42761–42768.

33 Xu, K., von Cresce, A., and Lee, U. (2010). *Langmuir* 26: 11538–11543.

34 Holoubek, J., Liu, H., Wu, Z. et al. (2021). *Nature Energy* 2021.

35 Xu, K. and von Wald Cresce, A. (2012). *Journal of Materials Research* 27: 2327–2341.

36 Thenuwara, A.C., Shetty, P.P., and McDowell, M.T. (2019). *Nano Letters* 19: 8664–8672.

37 Sazhin, S.V., Khimchenko, M.Y., and Tritenichenko, Y.N. (2000). *Journal of Power Sources* 87: 112–117.

38 Kulova Tatiana, L. (2020). *International Journal of Electrochemical Science*: 8638–8661.

39 Smart, C.M. (1999). *Journal of the Electrochemical Society* 146: 486–492.

40 Nobili, F., Dsoke, S., Mecozzi, T. et al. (2005). *Electrochimica Acta* 51: 536–544.

41 Zhang, S.S., Xu, K., Allen, J.L. et al. (2002). *Journal of Power Sources* 110: 216–221.

42 Ejp, A. and Wkb, B. (2000). *Journal of Power Sources* 88: 192–196.

43 Mandal, B.K., Padhi, A.K., Shi, Z. et al. (2006). *Journal of Power Sources* 162: 690–695.

44 Liao, X.-Z., Ma, Z.-F., Gong, Q. et al. (2008). *Electrochemistry Communications* 10: 691–694.

45 Smart, M.C., Ratnakumar, B.V., Whitcanack, L.D. et al. (2003). *Journal of Power Sources* 119–121: 349–358.

46 Smart, M.C., Ratnakumar, B.V., Whitcanack, L.D. et al. (2008). *ECS Transactions* 11.

47 Yang, L., Furczon, M.M., Xiao, A. et al. (2010). *Journal of Power Sources* 195: 1698–1705.

48 Fu, M.H., Huang, K.L., Liu, S.Q. et al. (2010). *Journal of Power Sources* 195: 862–866.

49 Sheng, S.Z. (2006). *Electrochemistry Communications* 8: 1423–1428.

50 McOwen, D.W., Seo, D.M., Borodin, O. et al. (2014). *Energy and Environmental Science* 7: 416–426.

51 Dong, X., Guo, Z., Guo, Z. et al. (2018). *Joule* 2: 902–913.

52 Dong, X., Lin, Y., Li, P. et al. (2019). *Angewandte Chemie International Edition.*

53 Hongming, Z., Xiao, K., and Li, J. (2016). *Journal of Power Sources* 302: 274–282.

54 Qiuping, Z., Yu, Z., Tang, F. et al. (2017). *Journal of the Electrochemical Society* 164: A1873–A1880.

55 Shangguan, X., Xu, G., Cui, Z. et al. (2019). *Small* 15: e1900269.

56 Thenuwara, A.C., Shetty, P.P., Kondekar, N. et al. (2020). *ACS Energy Letters* 5: 2411–2420.

57 Zheng, T., Xiong, J., Zhu, B. et al. (2021). *Journal of Materials Chemistry A* 9: 9307–9318.

58 Shui, Z.S. (2006). *Journal of Power Sources* 162: 1379–1394.

59 Herreyre, S., Huchet, O., and Barusseau, S. (2001). *Journal of Power Sources* 97: 576–580.

60 Smart, M.C., Lucht, B.L., Dalavi, S. et al. (2012). *Journal of the Electrochemical Society* 159: A739–A751.

61 Ralf, W., Brox, S., Kasnatscheew, J. et al. (2014). *Electrochemistry Communications* 40: 80–83.

62 Yang, B., Zhang, H., Yu, L. et al. (2016). *Electrochimica Acta* 221: 107–114.

63 Zhonghui, S., Zheng, L., Gao, L. et al. (2018). *Advanced Energy Materials* 9: 1802946.

64 Smart, M.C., Ratnakumar, B.V., Behar, A. et al. (2007). *Journal of Power Sources* 165: 535–543.

65 Gao, Y., Rojas, T., Wang, K. et al. (2020). *Nature Energy* 5: 534–542.

66 Yang, Y., Davies, D.M., Yin, Y. et al. (2019). *Joule* 3: 1986–2000.

67 Rustomji, C.S., Yang, Y., Kim, T.K. et al. (2017). *Science* 356.

68 Yang, Y., Yin, Y., Davies, D.M. et al. (2020). *Energy and Environmental Science* 13: 2209–2219.

69 Ma, S., Jiang, M., Tao, P. et al. (2018). *Progress in Natural Science: Materials International* 28: 653–666.

70 Wang, J., Huang, W., Pei, A. et al. (2019). *Nature Energy* 4: 664–670.

71 Thomas, W., Wilka, M., Kasper, M. et al. (2014). *Journal of Power Sources* 262: 129–135.

72 Haran, B., White, R., and Popov, B.N. (2002). *Journal of Power Sources.*

73 Striebel, K.A., Shim, J., Cairns, E.J. et al. (2004). *Journal of the Electrochemical Society* 151: A857.

74 Naoaki, Y. and Ohzuku, T. (2005). *Journal of Power Sources* 146: 636–639.

75 Elena, M., Pollak, E., Salitra, G. et al. (2007). *Journal of Power Sources* 174: 1263–1269.

76 Lucille, B., Dedryvère, R., Martinez, H. et al. (2012). *Journal of the Electrochemical Society* 159: A1739–A1746.

77 Madeleine, E., Gerschler, J.B., Vogel, J. et al. (2012). *Journal of Power Sources* 215: 248–257.

78 Araki, K. and Sato, N. (2003). *Journal of Power Sources* 124: 124–132.

79 Xiong, D.J., Petibon, R., Nie, M. et al. (2016). *Journal of the Electrochemical Society* 163: A546–A551.

80 Ma, S. and Noguchi, H. (2006). *Journal of Power Sources* 161: 1297–1301.

81 Kojima, Y., Muto, S., Tatsumi, K. et al. (2011). *Journal of Power Sources* 196: 7721–7727.

82 Heike, G., Ozawa, Y., and Yazami, R. (2006). *Electrochimica Acta* 52: 1499–1506.

83 Lucille, B., Naturel, R., Martinez, H. et al. (2013). *Journal of Power Sources* 236: 265–275.

84 Lydia, T., Weber, S., Blaske, F. et al. (2013). *Journal of Power Sources* 242: 832–837.

85 Xiulin, F., Ji, X., Chen, L. et al. (2019). *Nature Energy* 4: 882–890.
86 Jie, L., Xie, K., Lai, Y. et al. (2010). *Journal of Power Sources* 195: 5344–5350.
87 Hongbo, H., Guo, J., Zhang, D. et al. (2011). *Electrochemistry Communications* 13: 265–268.
88 Markovsky, B., Rodkin, A., Cohen, Y.S. et al. (2003). *Journal of Power Sources* 119–121: 504–510.
89 Tae-Heum, N., Shim, E.-G., Kim, J.-G. et al. (2008). *Journal of Power Sources* 180: 561–567.
90 Chia-Chin, C., Lee, K.-Y., Lee, H.-Y. et al. (2012). *Journal of Power Sources* 217: 524–529.

6
High-Concentration Electrolytes

6.1 High-Concentration Electrolytes

It is known that electrolytes play a vital role in the stable operation of batteries [1, 2]. In the past 30 years, $LiPF_6$/carbonate-based electrolyte has been widely used in Li-ion batteries (LIBs), due to the formation of a stable solid electrolyte interface phase (SEI) layer on the graphite surface, which has greatly improved the cycle stability of the batteries [3–5]. However, in Li metal batteries, the traditional $LiPF_6$/carbonate-based electrolyte does not exhibit excellent performance [6, 7]. This is because Li metal batteries cannot form a good SEI in traditional electrolytes, leading to uncontrollable growth of Li dendrites and low coulombic efficiency (CE) [8, 9]. $LiPF_6$ can react to form AlF_3 film on the anode current collector made up of aluminum foil, which effectively inhibits the electrolyte from corroding the aluminum current collector [10, 11]. However, when there is a small amount of water or at a higher temperature, $LiPF_6$ will be severely decomposed, resulting in a low electrochemical window of the electrolyte, which limits the realization of its high energy density [12, 13]. Although electrolyte additives have made a significant contribution to the improvement of the performance of lithium metal batteries (LMBs) [14–16], they still have problems such as high flammability, poor thermal stability, narrow electrochemical window, and slow electrode reaction kinetics, which seriously hinder their performance [17–19]. Therefore, there is an urgent need to develop a new type of high-safety electrolyte system. In this case, the strategy of high-concentration electrolyte (HCE) was successfully developed and applied in Li metal batteries.

6.1.1 Concept, Design Strategies

In nonaqueous solutions, usually when the concentration of Li salt is $\geq 3.0\,\mathrm{mol\,l^{-1}}$, a new type of electrolyte with a special three-dimensional (3D) solution structure is formed, called HCEs [20–22]. In HCEs, as the salt concentration continues to increase, the interaction between the cations and anions in the salt and the solvent is enhanced, and the content of free solvent molecules is greatly reduced [23, 24]. Therefore, HCEs exhibit unique physicochemical and electrochemical properties that are significantly different from those of conventional dilute electrolytes [25, 26].

Liquid Electrolyte Chemistry for Lithium Metal Batteries: Design, Mechanisms, Strategies,
First Edition. Jianmin Ma.

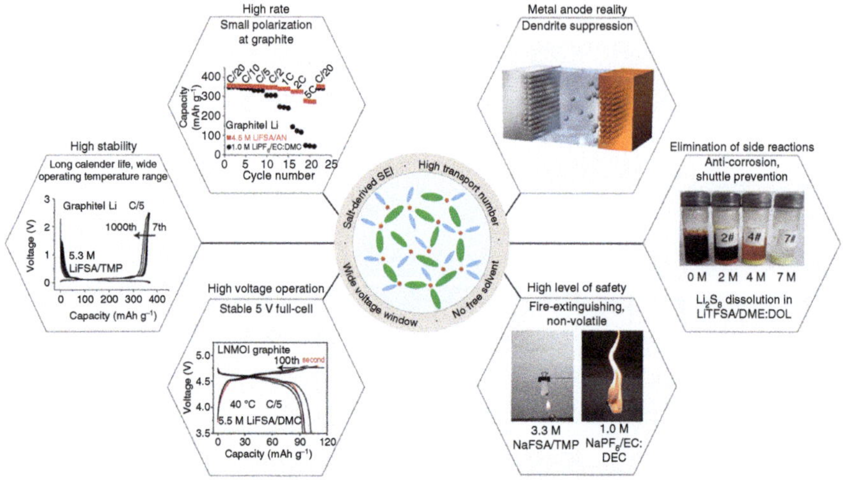

Figure 6.1 The central circle (pale blue) encloses a representative structure of a concentrated electrolyte surrounded by an outer circle containing the essential material properties (pale brown) that determine the physicochemical and electrochemical properties. Source: Yamada et al. [17]/with permission of Springer Nature. The outer hexagons list the various advanced battery functions: high rate, high stability, high-voltage operation, high level of safety, metal anode reality, and elimination of side reactions with LNMO.

In Table 6.1, the advantages and disadvantages of conventional dilute electrolyte systems and HCE systems are briefly compared [18]. Although high viscosity, low wettability, low ion conductivity, and high cost of HCEs restrict their further application, their unique properties still receive widespread attention [27–29]. In HCEs, the main components in the electrolyte are contact ion pairs (CIP) and cation–anion aggregates (AGG), in which the solvent molecules become completely coordinated, there are almost no free solvent molecules, and the anions are dominant in the solvation sheath [20, 24, 30]. As the anions are coordinated with multiple Li ions, the position of the lowest unoccupied molecular orbital (LUMO) moves from the position of the solvent to the position of the salt. Therefore, the anions in the electrolyte preferentially decompose at a low potential to form an anion-derived SEI with LiF-rich and exhibit higher mechanical strength and improved cycle life [20, 31]. The high-concentration salt strategy has improved the thermal stability of the electrolyte and reduced the flammability, alleviated the hidden safety hazards in traditional diluted electrolytes, and opened up a new path for the electrolyte project [25, 32, 33]. As shown in Figure 6.1 [17], the HCE has a salt-derived SEI, a wide voltage window, no free solvent molecules, and a high transmission number. Therefore, the battery can achieve high rate, high stability, high voltage window, high safety, eliminate side reactions, and inhibit the growth of dendrites [2, 17, 34].

6.1.2 Developments

Traditional commercial electrolytes are usually mixed solvents based on EC, mainly because EC can be reversibly embedded and extracted on the graphite electrode

Table 6.1 Comparison of conventional dilute electrolytes and HCE systems.

Physicochemical property	Dilute electrolytes	Concentrated electrolytes
Representative components in the bulk electrolyte	Solvent-separated ion pairs (SSIP) and free solvent molecules	Contact ion pairs (CIPs) and cation–anion aggregates (AGGs)
Representative components of the SEI	A combination of inorganic salts decomposed from the solute and organic species derived from solvents	Dominated by inorganic species from decomposed anions or from partially precipitated solutes
Flammability	High	Low
Thermal stability	Poor	Good
Reductive stability	Low	High
Oxidative stability	Low	High
Viscosity	Low	High
Ionic conductivity	10^{-2} S cm^{-1}	10^{-3} to 10^{-2} S cm^{-1}
Wettability	Good	Relatively poor
Electrode reaction kinetics	Slow	Fast
Power density	High	May exceed conventional $LiPF_6$/EC-based electrolytes at certain circumstances
Energy density	High	High
Cost	Low	High

Source: Zheng et al. [18]. Licensed under CC BY 4.0.

[35, 36]. Other organic solvents can easily cointercalate the graphite electrode to destroy the layered structure and irreversibly store Li [37, 38]. Limited by the requirements of solvent composition, organic electrolytes have made little progress, but the emergence of HCEs broke this convention [6]. HCEs were first used to stabilize graphite/Li metal cells (Gr/Li). Jeong et al. in 2003 [39] found for the first time that HCEs can passivate graphite electrodes in cells, showing completely different performance from diluted electrolytes. As the research further deepens, high-concentration $LiN(SO_2C_2F_5)_2$/PC (3.27 mol kg^{-1}) was used as an electrolyte to inhibit the growth of Li dendrites [40]. As shown in Figure 6.2a,c, the thickness of the SEI layer in the 1.28 mol kg^{-1} electrolyte is ~35 nm, whereas it is even thinner, ~20 nm in the 3.27 mol kg^{-1} electrolyte (Figure 6.2b,d). This shows that in the HCE, dense SEI is formed and the formation of Li dendrites is inhibited. It is worth noting that this work has set a precedent for the application of HCEs in batteries. Although high-concentration Li salts exhibited excellent performance in PC solvents, the detailed mechanism of electrode–electrolyte interface formation in electrolytes

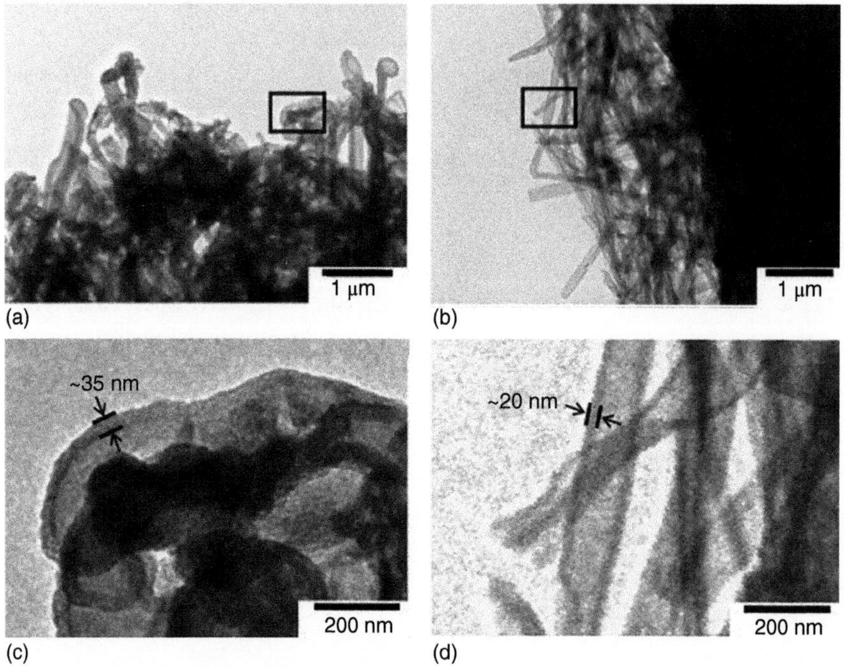

Figure 6.2 TEM images of the nickel substrates obtained after the 30th cycle in $LiN(SO_2C_2F_5)_2$/PC (a,c) 1.28 mol kg^{-1} and (b,d) 3.27 mol kg^{-1}. Source: Jeong et al. [40], Figure 02/with permission of Elsevier.

of different concentrations was not elucidated at that time. Subsequently, Nie et al. conducted an in-depth exploration of the behavior of the battery in different concentrations of $LiPF_6$/PC electrolytes [22]. It can be seen from Figure 6.3a,b that as the Li salt concentration increases, the Li intercalation and deintercalation behavior of the graphite electrode changes. In the diluted concentration $LiPF_6$/PC (1.2 M), the graphite electrode exhibits continuous electrolyte reduction and no lithiation/delithiation of the graphite. When the high concentration of $LiPF_6$ in PC (3.0–3.5 M), charging and discharging behavior reveals reversible lithiation/delithiation of the graphite electrode. EDX and TEM images also show that the electrolyte concentration is closely related to the electrode–electrolyte interface (Figure 6.3c,d). In the diluted concentration electrolyte, the electrode surface film is sparse, unable to passivate the graphite electrode, resulting in continuous electrolyte reduction. On the contrary, in the HCEs, the surface film coats the graphite particles uniformly and densely, which inhibits the continuous reaction of the electrolyte. In addition, EDX images (the insets in Figure 6.3c,d) show that the element content of F and O in the graphite surface film in the HCE is much higher than that in the diluted concentration electrolyte. This is because the main structure in the HCE is the CIP. The CIP forms a LiF-rich surface film during the reduction process, which can passivate the electrode surface and realize reversible lithiation/delithiation of the graphite electrode [30, 41].

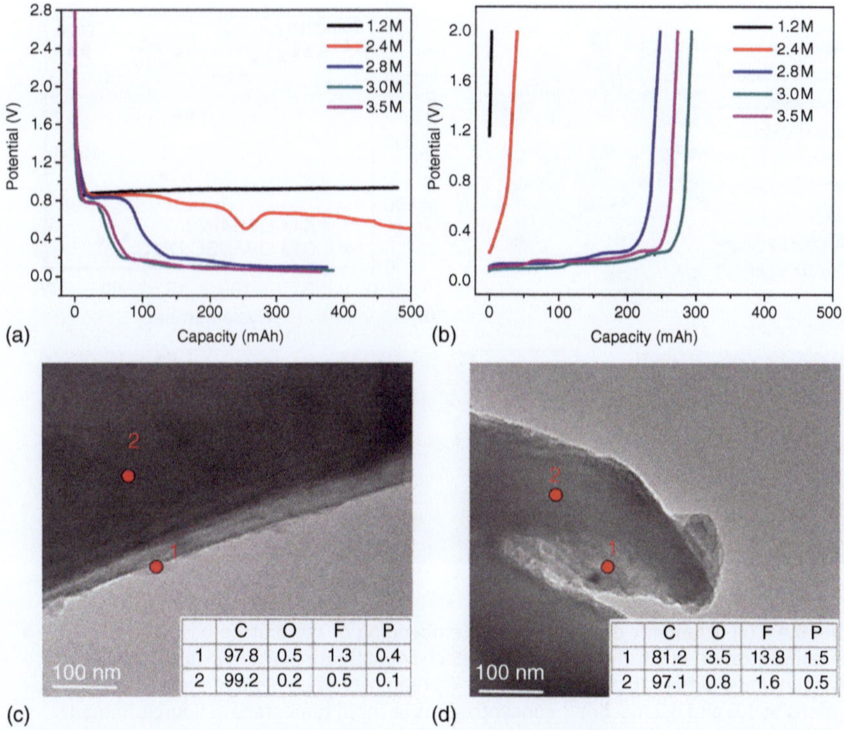

Figure 6.3 Potential vs. capacity curves for BF-graphite/Li cells cycled with different concentrations of $LiPF_6$/PC electrolytes: (a) Li intercalation profiles and (b) Li deintercalation profiles. TEM images of graphite anodes cycled with (c) 1.2 and (d) 3.5 M. The inset indicates the element composition detected by EDX. The red spots indicate locations probed by EDX. Source: Nie et al. [22]/with permission of American Chemical Society.

In addition to PC, organic solvents such as acetonitrile (AN) [20], sulfolane (SL) [31], dimethyl sulfoxide (DMSO) [31], tetrahydrofuran (THF) [31], dimethyl ether (DME) [42], have all been considered for use in batteries through HCEs. In the work of Yamada et al. [20], for the first time, AN was used as a solvent for HCEs in batteries. As shown in Figure 6.4a,b, the reversible insertion of Li^+ into graphite electrodes was successfully achieved through the HCE strategy in reduction-vulnerable AN solvent, and the cells using 4.5 M LiFSI/AN electrolyte at different current densities show much higher rate performance than the traditional diluted electrolyte. This is because the anion-derived SEI layer in the HCE has very unique characteristics and can realize rapid Li^+ transfer. In addition, the stability of the electrolyte can be seen from the reactivity of the Li metal and the electrolyte of different concentrations of Li salt (Figure 6.4c). After one week, it can be seen that the Li metal foil is very unstable in the diluted electrolyte, which causes the electrolyte to turn yellow. In sharp contrast, in the HCE, neither the Li metal nor the electrolyte has obvious changes, which indicates that the HCE overcomes the inherently poor reduction stability of the AN solvent, thereby improving the stability of the electrolyte.

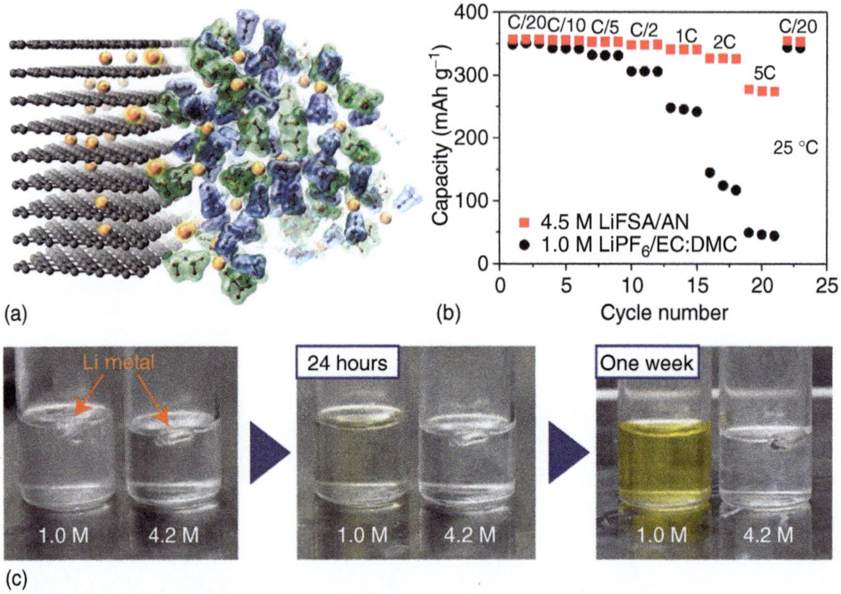

Figure 6.4 (a) Schematic diagram of rapid embedding of acetonitrile-based high-concentration electrolyte in graphite electrode; (b) rate performance of graphite/Li metal cell under different current densities; (c) reactivity of Li metal foil and LiTFSA/AN solutions at 1.0 and 4.2 mol dm^{-3} concentrations at room temperature. Source: Yamada et al. [20]/with permission of American Chemical Society.

In subsequent work, Yamada et al. proved that in HCEs composed of ether, sulfoxide, and sulfone organic solvents, Li ions can be reversibly inserted into graphite electrodes [31]. As shown in Figure 6.5a, the solvent will cointercalate Li ions into the layered structure of the graphite electrode, resulting in destruction of the graphite structure in the traditional diluted electrolyte [43]. Figure 6.5b shows that the cointercalation of Li solvent is inhibited in the HCE, and the Li ion is reversibly intercalated into the graphite electrode. It can be seen from the Raman spectrum in Figure 6.5c that there is a large amount of free DMSO molecules in the 1.0 M electrolyte, and both free DMSO molecules and solvated DMSO molecules are present in this dilute solution. In the 3.2 M electrolyte, only the coordination peak of Li^+-DMSO appeared, indicating that there are no free DMSO molecules, so Li ions can be highly reversibly inserted into the graphite electrode. This work breaks the traditional idea that LIBs can only use EC-based electrolytes and opens the way for advanced LIBs based on various EC-free organic electrolytes with various characteristics.

Li salt, as a vital component in the electrolyte, is also a factor that cannot be ignored [44]. $LiPF_6$ is a Li salt with relatively poor chemical and thermal stability [45–47]. HF generated by a chemical reaction with trace water or H^+ can cause the transition metal in the positive electrode material to dissolve and corrode the electrode material [48–51]. To solve this problem, Liu et al. inhibited the dissolution of transition metals by using a HCE of 3 M $LIPF_6$-EC/ethyl methyl carbonate (EMC)/DM [52]. As shown in Figure 6.6a,b, after 250 cycles, the cycle performance of the traditional

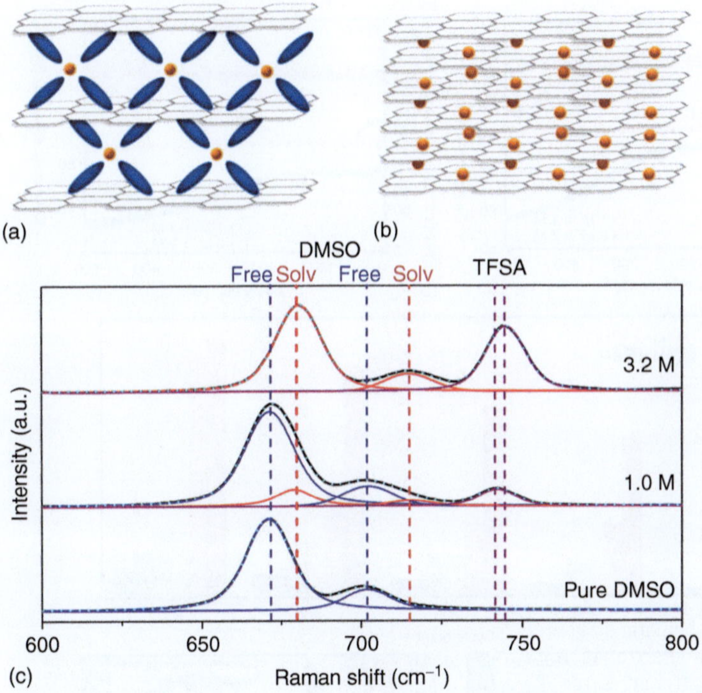

Figure 6.5 Images of (a) solvent cointercalation and (b) Li^+ intercalation; (c) Raman spectra of LiTFSI/DMSO solutions at various salt concentrations. Source: Reproduced with permission Yamada et al. [31].

diluted electrolyte drops rapidly at 2 and 5 C current densities. In sharp contrast, after 500 cycles, the HCE can still maintain the discharge specific capacity of 150 and 125 mAh g^{-1}, and the capacity retention rate is 94% and 100%, respectively. It is worth noting that even at a high current density of 5 C, the HCE can still maintain excellent electrochemical performance. This indicates that the LMB in the HCE has not only excellent cycle stability but also excellent rate performance, which is conducive to the rapid charge and discharge of the battery. In addition, the transition metal deposition diagram in different cycles (Figure 6.6c) reflects that as the number of cycles increases, the transition metal deposition in the traditional electrolyte gradually increases, while the transition metal deposition in the HCE only slightly increases, which further proves that the HCE can effectively inhibit the dissolution of transition metals. The schematic diagram of the formation of the cathode electrolyte interphase (CEI) in different electrolytes (Figure 6.6d) shows that the CEI in the traditional electrolyte is composed of decomposed solvent molecules with high organic content, showing the characteristics of fragility and unevenness [53]. Therefore, the electrode material cannot be effectively protected, and a large amount of the transition metal is dissolved and deposited on the electrode surface, resulting in poor cycle stability and speed performance and low CE. Compared with commercial electrolytes, $LiPF_6$ electrolyte in a high concentration tend to decompose and form the

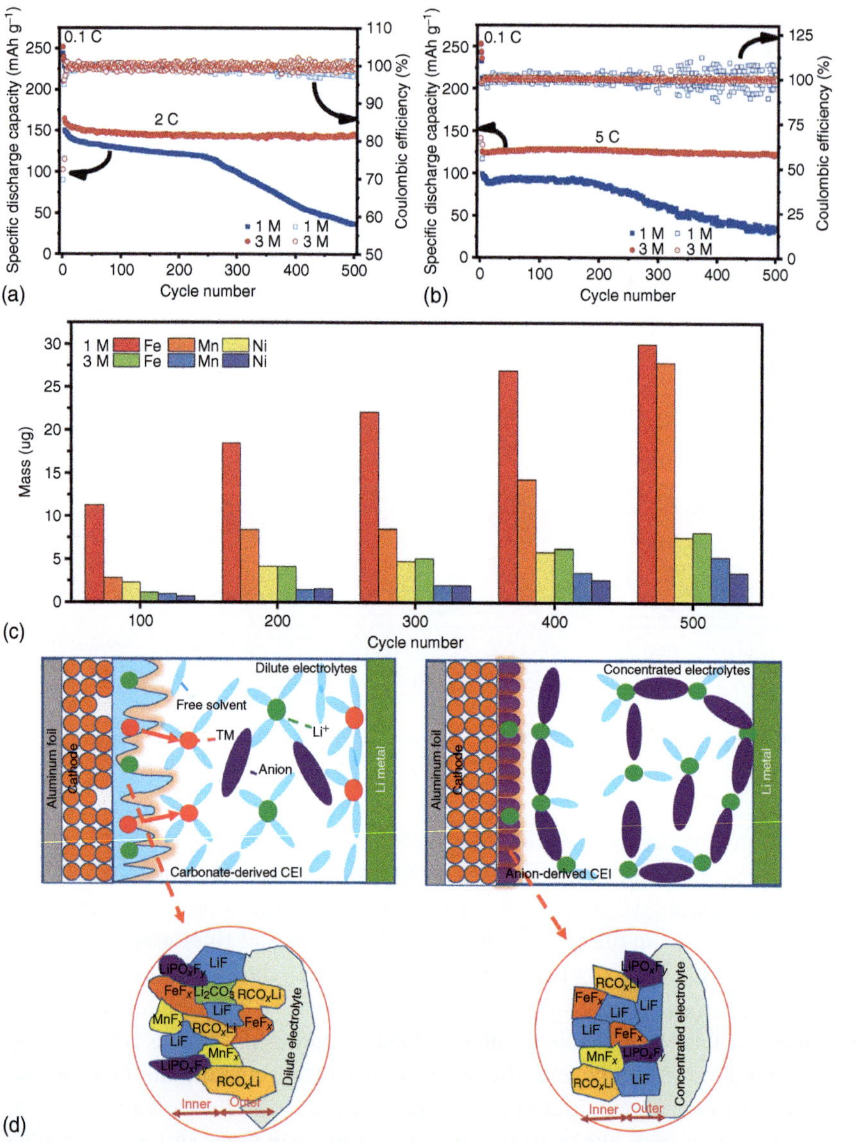

Figure 6.6 The cycle performance in 1 and 3 M electrolytes at 2 C (a) and 5 C (b); (c) the amount of transition metal deposited on different cycled Li plates in 1 and 3 M electrolytes at 2 C; (d) schematic illustrations of passivation films in 1 and 3 M electrolytes. Source: Reproduced with permission Liu et al. [52].

LiF-rich CEI. The LiF-rich CEI exhibits strong, uniform, and dense characteristics, which can effectively protect electrode materials and inhibit the dissolution of transition metals [16, 54], thereby exhibiting excellent electrochemical performance.

LiFSI and LiTFSI are Li salts with higher chemical and thermal stability than $LiPF_6$, so huge attention has been paid to electrolytes [55–57]. However, the

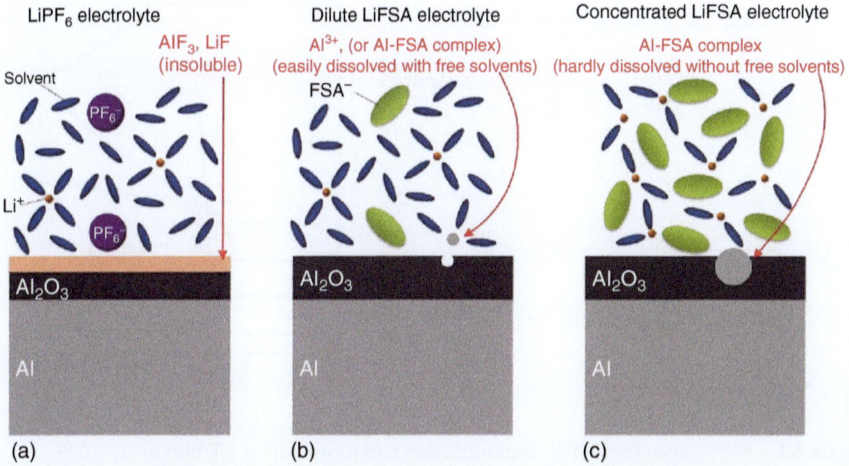

Figure 6.7 Schematic illustrations of the behavior of Al current collector in (a) conventional $LiPF_6$ electrolyte, (b) dilute LiFSI/AN electrolyte, and (c) highly concentrated LiFSI/AN electrolyte. Source: Reproduced with permission Yamada et al. [61].

LiFSI/LiTFSI salt cannot passivate the Al current collector during the charging process of conventional electrolytes, resulting in corrosion of the Al current collector, especially under high voltage conditions [57–60]. Therefore, in order to apply LiFSI and LiTFSI on a large scale, corrosion of the Al current collector must be overcome. In recent years, the strategy of using HCEs has effectively resolved Al current collector corrosion. In the work of Yamada and his collaborators, a high-concentration LiTFSI/AN electrolyte is used to inhibit corrosion of the Al current collector at a high voltage of 4.5 V [61]. The schematic diagram in Figure 6.7 shows that in the traditional diluted $LiPF_6$ electrolyte, the F^- anion produced by the hydrolysis of PF_6^- combines with Al^{3+} to form a thin layer of AlF_3 (or combines with Li^+ to form LiF) on the outer layer of Al_2O_3. Due to the strong ionic bond between AlF_3 and LiF, it is difficult to dissolve in organic electrolytes [60, 61]. Even at a high potential, corrosion of Al current collector is significantly suppressed. In the diluted LiFSI electrolyte, due to higher thermal stability, less F^- anions are produced, and Al^{3+} can only be combined with the solvent to form $Al(FSI)_3$. The free AN solvents, with strong solvation ability, can be dissociated, can be dissociated and dissolved $Al(FAI)_3$ complex by solvation [61–63], and then corrode the surface Al_2O_3 and the intimal Al metal (Figure 6.7b). The difference is that in HCEs, because there are no free AN solvent molecules and a large amount of FSI^-, the dissolving ability of $Al(FSI)_3$ complex is extremely low, and the dissolution of Al^{3+} is quite slow [61, 63, 64]. This increases the concentration of Al^{3+} at the interface, which ultimately leads to the oxidative decomposition of Al_2O_3 and Al metal, requiring a higher electrode potential (Figure 6.7c). Therefore, the protection of the Al current collector can be realized even at a high pressure of 4.5 V in a highly concentrated electrolyte.

Aiming at the corrosion problem of the Al current collector in the LiTFSI electrolyte, Matsumoto et al. achieved the purpose of protecting the Al current collector

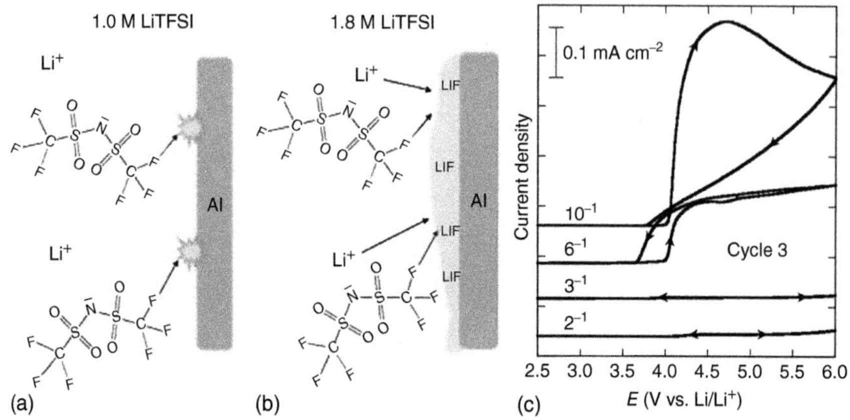

Figure 6.8 Mechanism for LiTFSI decomposition and formation of LiF film in (a) 1.0 M LiTFSI and (b) 1.8 M LiTFSI. Source: Reproduced with permission Matsumoto et al. [65]. (c) Third cycle from CVs of coin cells with Al foil working electrodes and Li metal counter/reference electrodes. Source: Reproduced with permission McOwen et al. [24].

by adjusting the concentration of Li salt (Figure 6.8a,b) [65]. As the concentration increases, LiTFSI is more likely to decompose to form LiF, thereby forming a thick LiF layer on the electrode surface, further inhibiting the corrosion of the Al current collector [60, 65, 66]. Henderson et al. believed that the $TFSI^-$ anions in the diluted electrolyte have a strong C—F bond and cannot be oxidized to form the F^- anions [24]. Therefore, the Al current collector cannot be passivated in the diluted LiTFSI/EC electrolyte. In contrast, for HCEs, the anodic stability of the Al surface is improved due to the extensive coordination of the electron lone pairs on the solvent and the anions with the positively charged Li^+ cations (Figure 6.8c) [24]. Due to the extensive coordination of the solvent molecules and anions, the solubility of the $Al(TFSI)_3$ complex in the concentrated electrolyte is low. The high concentration of $TFSI^-$ anions at the electrode/electrolyte interface acts as a barrier to prevent the solvent molecules from entering the electrode, further preventing Al from dissolving from the electrode surface.

Generally, ether-based electrolytes have better compatibility with Li metal than carbonate-based electrolytes [67–69], but dendrites still form on Li metal during the cycle. Qian et al. designed a high-concentration LiFSI/DME electrolyte to achieve high rate and long-cycle stability for LIBs [70]. As shown in Figure 6.9a,b, there are a large number of free solvated molecules (light gray areas) in the diluted electrolyte, so the uncoordinated solvent molecules are likely to react with Li metal and reduce CE [71]. In HCEs, there are no free solvent molecules, and almost all ions exist in the form of CIPs and AGGs. This effectively stabilizes the solvent molecules and reduces the reaction of the solvent with Li metal. In addition, a dense and highly conductive SEI layer is formed on the surface of the Li metal, thereby reducing the decomposition of anions during long cycles. Therefore, HCEs exhibit good cycle stability during the circulation process rather than the performance degradation shown in the diluted electrolyte (Figure 6.9c,d). It is worth noting that even at a high current

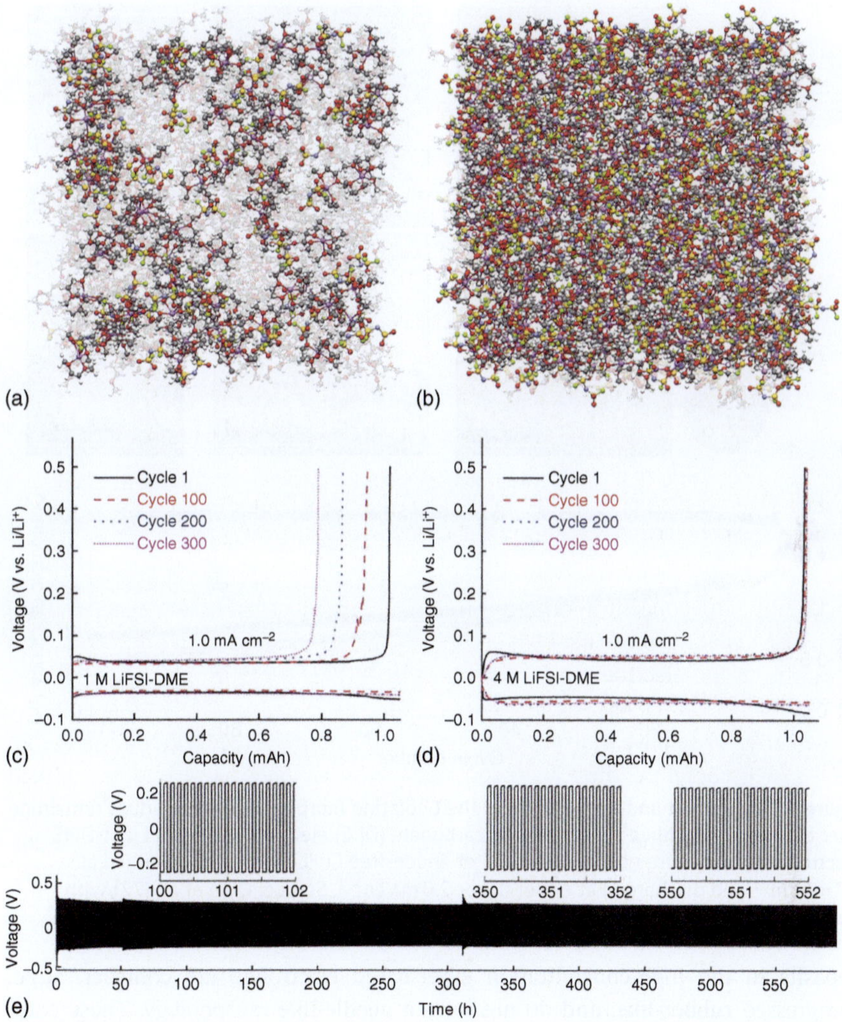

Figure 6.9 Snapshots of the MD simulation boxes (a) 1 M LiFSI-DME electrolyte and (b) 4 M LiFSI-DME electrolyte. The uncoordinated DME solvent molecules are colored light gray; voltage profiles for the cell cycled in (c) 1 M LiFSI-DME and (d) 4 M LiFSI-DME electrolyte; (e) Li|Li cell cycled at 10.0 mA cm^{-2} with a 4 M LiFSI-DME electrolyte; the top plots are expanded views from the bottom plot. Source: Qian et al. [70]. Licensed under CC BY 4.0.

density of 10 mA cm^{-2}, HCEs can still maintain excellent cycle stability (Figure 6.9e). This is because of high reduction stability of the electrolyte and the high Li ion concentration enables high-rate Li plating/stripping.

In the work of Qian et al. a high-concentration ether-based electrolyte (4 M LiFSI/DME) was further used for anode-free Li batteries [72]. From the optical and SEM images (Figure 6.10a,b), it can be seen that in the diluted carbonate-based electrolyte, many needle-like dendrites of different shapes and sizes are formed at a relatively high current density, which can penetrate the diaphragm. Li

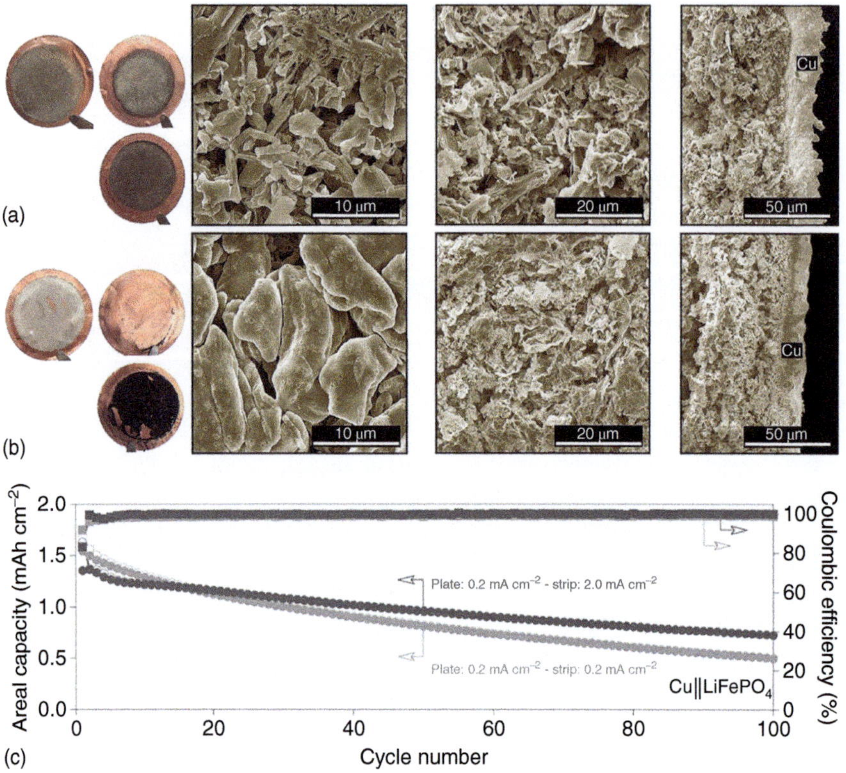

Figure 6.10 Optical and SEM images of the Li plating morphology and residues remaining after discharge (a) 1 M $LiPF_6$-EC/ diethyl carbonate (DEC) electrolyte, (b) 4 M LiFSI-DME electrolyte; (c) discharge capacity and CE of anode-free Cu||$LiFePO_4$ cells charged at 0.2 mA cm^{-2} and discharged at either 0.2 or 2.0 mA cm^{-2}. Source: Qian et al. [72]/with permission of WILEY-VCH.

deposits in the high-concentration ether-based electrolyte are compact, large, compressed rubber-like, and do not have a needle-like morphology. These compressed rubber-like Li deposits have a smaller surface area, which can effectively reduce the continuous reaction between Li and the electrolyte, thereby ensuring the sustainable operation of the anode-free Li battery. In addition, the discharge rate during the cycle can also effectively improve the capacity retention rate of the battery. As shown in Figure 6.10c, under the same charge rate, at a higher discharge rate of 2.0 mA cm^{-2}, the battery capacity retention rate is 54% of its original capacity after 100 cycles, which is much higher than the capacity retention rate (32%) at the 0.2 mA cm^{-2} discharge rate.

6.2 Local High-concentration Electrolytes

Compared to traditional diluted electrolytes, although HCEs have many advantages, there are some shortcomings that hinder their commercial development

[73, 74]. Currently, two main problems are restricting the development of HCEs: high viscosity and high cost [17, 19, 75]. Commercially, batteries using traditional electrolytes require 24 hours of wetting before use, whereas HCEs have a higher viscosity because they need to be soaked for longer before use [76, 77]. In addition, the main cost in the electrolyte comes from the Li salt, and the high concentration of the Li salt in HCEs is bound to bring high costs [17, 78]. Therefore, the local high-concentration electrolytes (LHCEs) strategy has been developed to solve the problems in HCEs.

6.2.1 Concept, Design Strategies

LHCEs introduce ideal diluents on the basis of HCEs, to reduce the concentration and cost [79, 80]. An ideal thinner should meet the following conditions: low cost, no additional cost; low viscosity and high wettability, effectively reducing the viscosity of the electrolyte; high safety, without increasing the risk of battery safety problems; low dielectric constant and low complexing ability, compatible with solvents, but not compatible with Li salts, does not change the local coordination environment and high solubility of HCEs; high chemical stability, to ensure that the electrochemical window is not shortened [17, 81, 82]. Figure 6.11 shows the general structure of three electrolytes. Unlike the traditional diluted electrolyte (Figure 6.11a) and the HCE (Figure 6.11b), in the LHCE, the introduced non-solvated diluent molecules will form a unique solvation structure, where the overall solvation characteristics of the HCE are retained in the inner solvation sheath, and the diluent forms an external non-solvent sphere (Figure 6.11c). Compared with the HCE, the non-solvated diluent outer sphere not only can improve the physical properties of the LHCE (for example, viscosity and wettability) but also can protect the cations and anions in the solvated sheath from the ions in the nearby solvated complex [79, 83]. As a result, the cation–anion and ion–solvent interactions in the internal solvation sheath of the LHCE are further enhanced.

6.2.2 Developments

Recently, in the work of Ren et al. 1,1,2,2-tetrafluoroethyl 2,2,3,3-tetrafluoropropyl ether (TTE) was added as a diluent to a highly concentrated LiFSI-TMS electrolyte

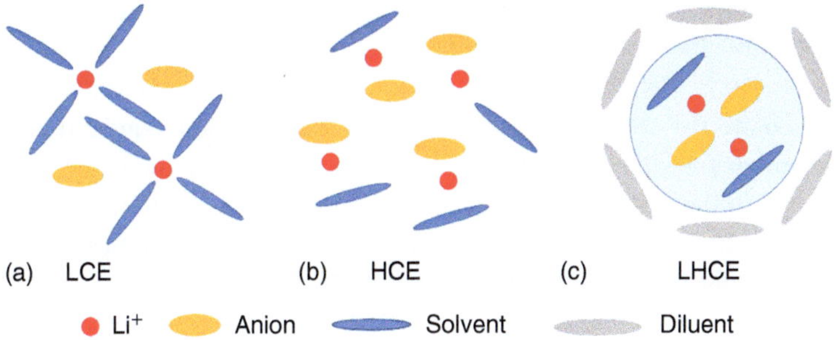

Figure 6.11 Schematic diagram of electrolyte structures (a) LCE, (b) HCE, and (c) LHCE.

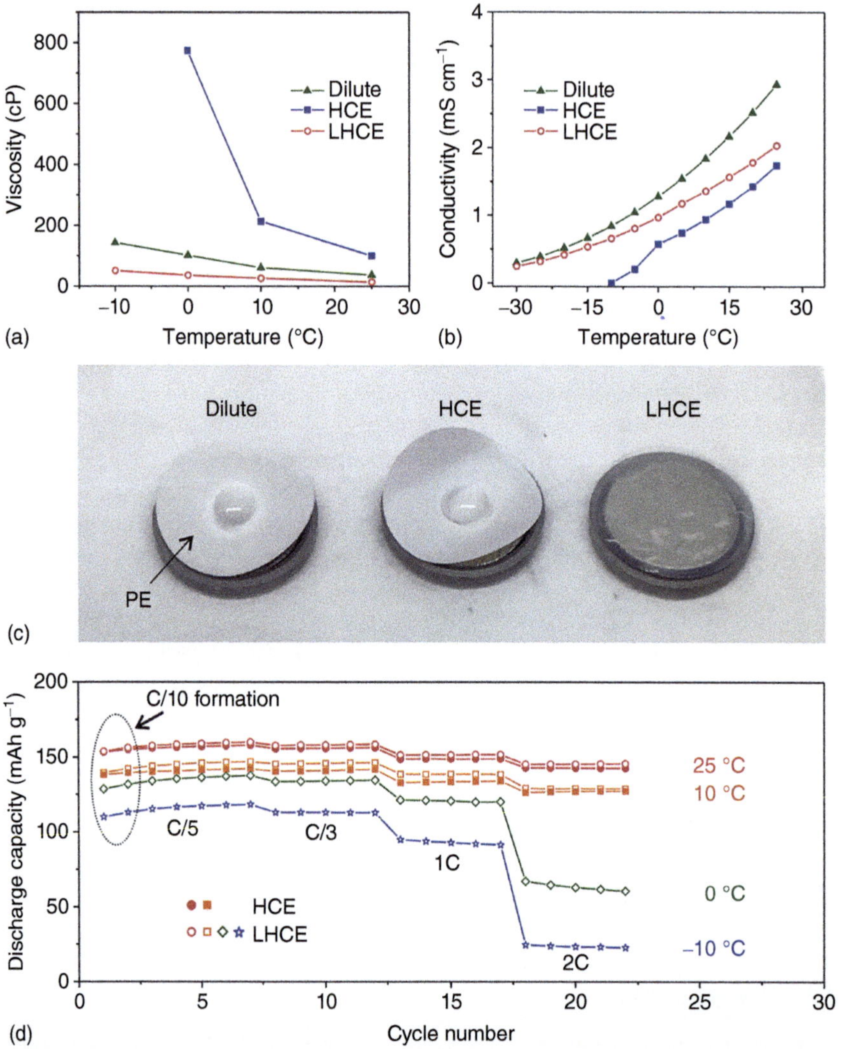

Figure 6.12 (a) Viscosity and (b) ionic conductivity of sulfone-based electrolytes under different temperatures; (c) wettability tests of sulfone electrolytes on polyethylene (PE) separator; (d) comparison of discharge rate capabilities in HCE and LHCE at different temperatures. Source: Ren et al. [84]/with permission of Elsevier.

[84]. Figure 6.12a,b are the change curves of the viscosity and ionic conductivity of the three electrolytes at different temperatures. It can be understood that the viscosity of the HCE in the test temperature range is always higher than that of the traditional diluted electrolyte, and the ion conductivity is also lower than that of the traditional diluted electrolyte. Remarkably, after introducing a thinner to obtain an LHCE, the viscosity and ionic conductivity of the electrolyte have drastically changed. The viscosity of the LHCE is much lower than that of the HCE, and even lower than that of the traditional diluted electrolyte. The ionic conductivity

of the LHCE is improved in a wide range. In comparison of the wettability of PE separator and electrolyte (Figure 6.12c), the LHCE can be easily dispersed and infiltrated on the separator, which is a phenomenon that does not appear in the other two electrolytes. In addition, in the magnification test at different temperatures (Figure 6.12d), the LHCE has shown excellent performance and can continue to run at 0 °C or even −10 °C.

In the subsequent work, Ren et al. added TTE to the high-concentration LiFSI-DME electrolyte and tested it under actual application conditions [85]. From the ^{17}O nuclear magnetic resonance (NMR) spectrum (Figure 6.13a), it can

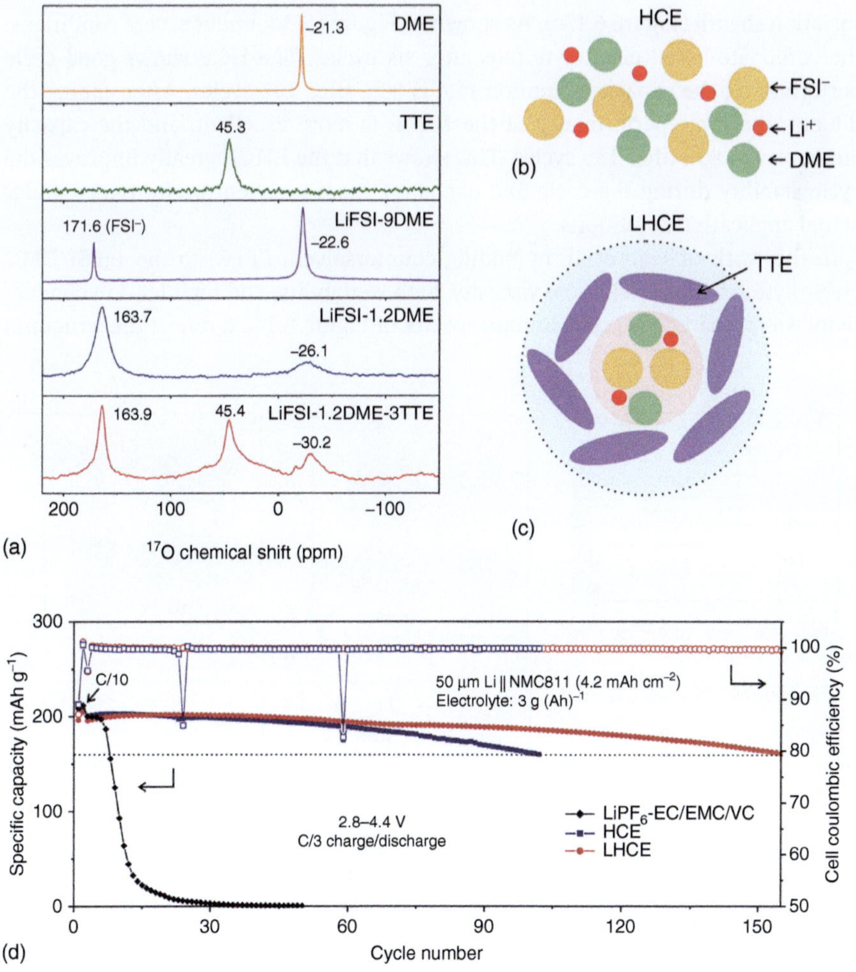

Figure 6.13 (a) ^{17}O-NMR spectra of different solvents and electrolytes; the scheme of the solvation structure in (b) HCE (LiFSI-1.2DME) and (c) LHCE (LiFSI-1.2DME-3TTE); (d) cycling performances of Li||NMC811 cells in different electrolytes with a 50 μm Li anode, 4.2 mAh cm^{-2} NMC811 cathode, and lean electrolyte at 14 μl. Source: Reproduced with permission Ren et al. [85].

be observed that as the electrolyte concentration increases, the chemical shift of sulfonyl oxygen decreases from 171.6 to 163.7 ppm, which indicates that the ionic dipole interaction between FSI^- and Li^+ in the HCE is enhanced. In addition, the oxygen signal in the ether state is also reduced because the DME in the solvation is replaced by FSI^-. The changes in these signals may be caused by the complex solvation structure (Figure 6.13b). With the addition of TTE, the chemical shifts of ether oxygen atoms are further reduced, while the oxygen signals from sulfonyl and TTE are almost unchanged, which indicates that the TTE diluent will not interrupt the close interaction between FSI^- and Li^+ in the HCE. On the contrary, more FSI^- anions entered the inner Li^+ solvation shell of the LHCE, which may be related to the different affinity of DME and FSI anions with TTE molecules outside the solvation sheath (Figure 6.13c). As shown in Figure 6.13d, under severe conditions, the carbonate-based electrolyte fails after six cycles. The HCE shows good cycle performance; the capacity retention rate is 80% after 102 cycles. After adding the diluent, the cycle performance of the LHCE is more excellent, and the capacity attenuates to 80% after 155 cycles. This shows that the LHCE greatly improves the cycle stability during the cycle and can show more excellent performance under actual application conditions.

In the work of Piao et al. by adding countersolvent (TTE) to the LiFSI/DMC electrolyte, an LHCE with low viscosity, high wettability, and high Li^+ ion conductivity was prepared [86]. The Raman spectra of Figure 6.14a,d reflect the structural

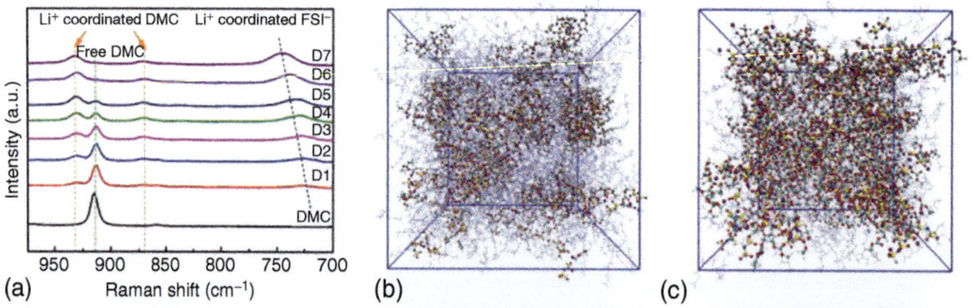

Figure 6.14 Raman spectra of LiFSI/DMC electrolytes in the range of 700–975 cm^{-1} (a) at different compositions (D1–D7), (d) TTE, D7, and T3; snapshots of different LiFSI/DMC electrolytes obtained by MD simulations at 298 K: (b) D1, (c) D4, (e) D7, and (f) T3. The Li^+ ions and their first coordinated shells (within 2.75 Å of Li^+ ions) are presented by ball-and-stick model, while the wireframes stand for free DMC (blue) and free TTE (green), respectively. Source: Reproduced with permission Piao et al. [86].

changes of LiFSI/DMC/TTE electrolytes with different compositions. With the increase in the Li salt concentration in the electrolyte, the free DMC molecular peaks gradually weakened, and instead, stronger Li^+-DMC solvated molecules appeared. When TTE was introduced, the coordination peak of Li^+-FSI^- in the electrolyte (LiFSI/DMC/TTE 1 : 1.5 : 1.5) shifted to the left, indicating that TTE can enhance the coordination strength of Li^+ and FSI^-. In addition, molecular dynamics (MD) simulations also confirmed that there is only a small amount of free DMC molecules in the electrolyte (LiFSI/DMC 1 : 7), indicating that the HCE is mainly a 3D network structure composed of Li^+-DMC and Li^+-FSI^-. In the LHCE (LiFSI/DMC/TTE 1 : 1.5 : 1.5), the structure is divided into independent local 3D structures by TTE, which proves that TTE not only does not directly solvate with Li^+ but can also enhance the solvation strength of Li^+ and FSI^-.

Recently, in the work of Chen et al. bis(2,2,2-trifluoroethyl) ether (BTFE) was used as a diluent for 3.2 M LiFSI/triethyl phosphate (TEP) electrolyte [87]. As shown in the schematic diagram of Figure 6.15a, BTFE does not dissolve Li salt in the LHCE and does not destroy the solvation structure of $Li_{(TEP)}$-FSI, so it can maintain its good electrochemical stability [88]. In addition, TEP is the best flame retardant [89, 90], and the LHCE also shows excellent flame retardancy in the ignition test (Figure 6.15b). In the Raman spectra (Figure 6.15c), compared with the HCE,

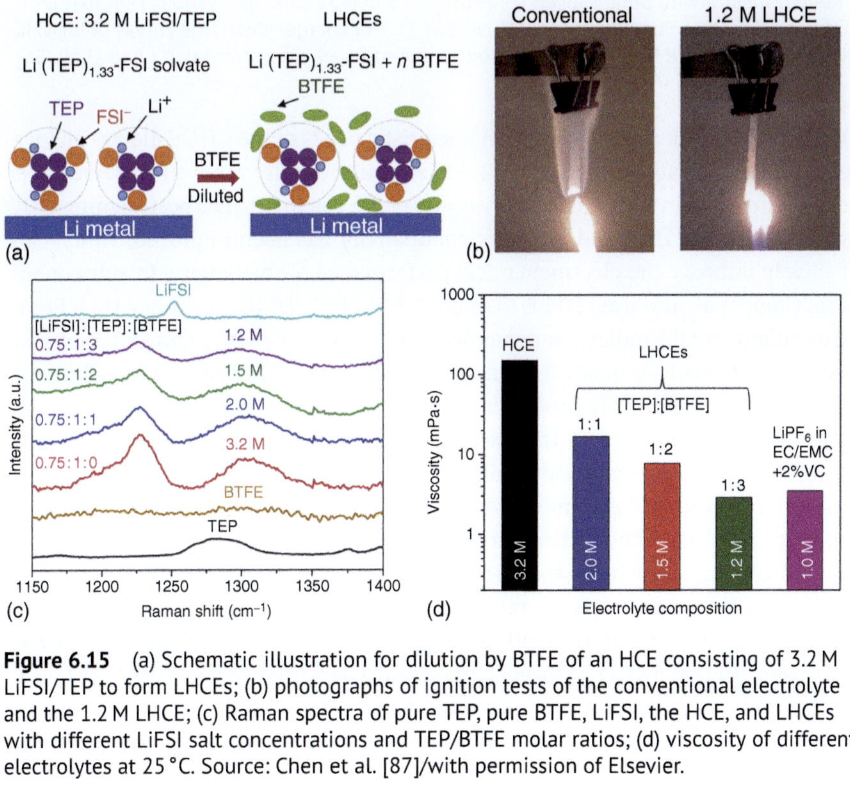

Figure 6.15 (a) Schematic illustration for dilution by BTFE of an HCE consisting of 3.2 M LiFSI/TEP to form LHCEs; (b) photographs of ignition tests of the conventional electrolyte and the 1.2 M LHCE; (c) Raman spectra of pure TEP, pure BTFE, LiFSI, the HCE, and LHCEs with different LiFSI salt concentrations and TEP/BTFE molar ratios; (d) viscosity of different electrolytes at 25 °C. Source: Chen et al. [87]/with permission of Elsevier.

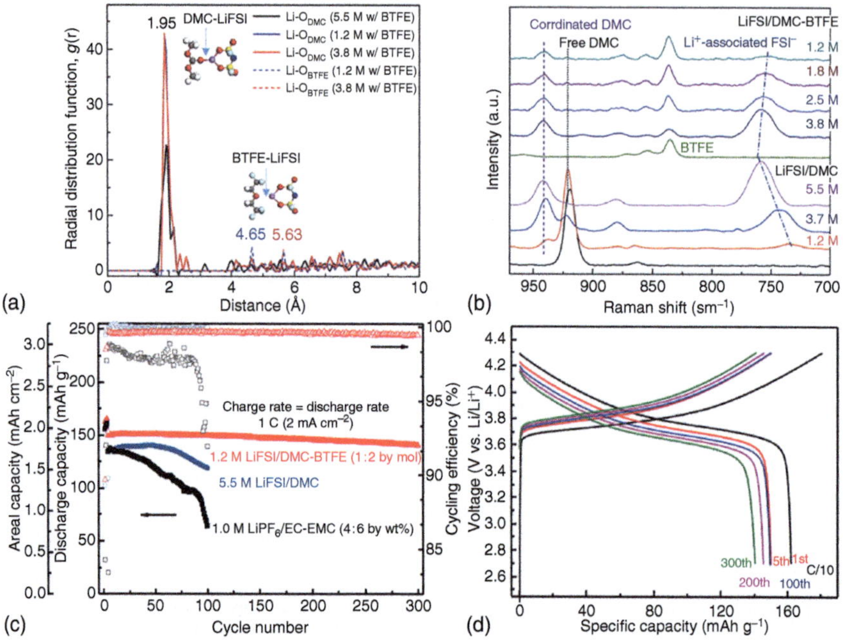

Figure 6.16 (a) Radial distribution functions of Li-O$_{DMC}$ and Li-O$_{BTFE}$ pairs; (b) progression of Raman spectra with different salt concentrations in pure DMC and various DMC/BTFE mixtures; (c) cycling stability of Li||NMC cells at 1 C; (d) charge–discharge curves of Li||NMC cells of 1.2 M LiFSI/DMC-BTFE at 0.1 C. Source: Reproduced with permission Chen et al. [81].

after the introduction of BTFE, the characteristic peaks of the LHCE did not change significantly, indicating that the Li$_{(TEP)}$-FSI solvent structure was well preserved. It is worth noting that in the LHCE, the viscosity of the electrolyte has been significantly reduced (Figure 6.15d), and the ionic conductivity has been improved, which can effectively improve the electrochemical performance of the battery. In subsequent work, Chen et al. also used BTFE to dilute 5.5 M LiFSI/DMC electrolyte [81]. From the calculation of the molecular dynamics simulation trajectory (Figure 6.16a), it can be seen that the solvation of Li^+ is mainly carried out by DMC molecules, while the interaction between BTFE and other electrolyte components is weak. In addition, it can be seen from Figure 6.16b that the characteristic peak of Li^+-FSI^- in 1.2 M LiFSI/DMC-BTFE is weakened, which is beneficial to promote the diffusion of Li ions and improve the ion conductivity and kinetic performance of the electrolyte. Therefore, the LHCE showed excellent performance in the cycle test. As shown in Figure 6.16c, the capacity of the traditional electrolyte decays rapidly at a current density of 1 C, and the capacity retention rate is only 40% after 100 cycles. In the HCE, although battery performance has improved, its capacity retention rate is 76%. After being diluted by BTFE, the electrochemical performance is further improved, and the cycle life is significantly increased. After 300 cycles, the discharge capacity is still stable at 150 mAh g^{-1} (capacity retention rate is 95%), and the average CE is 99.7% (Figure 6.16d).

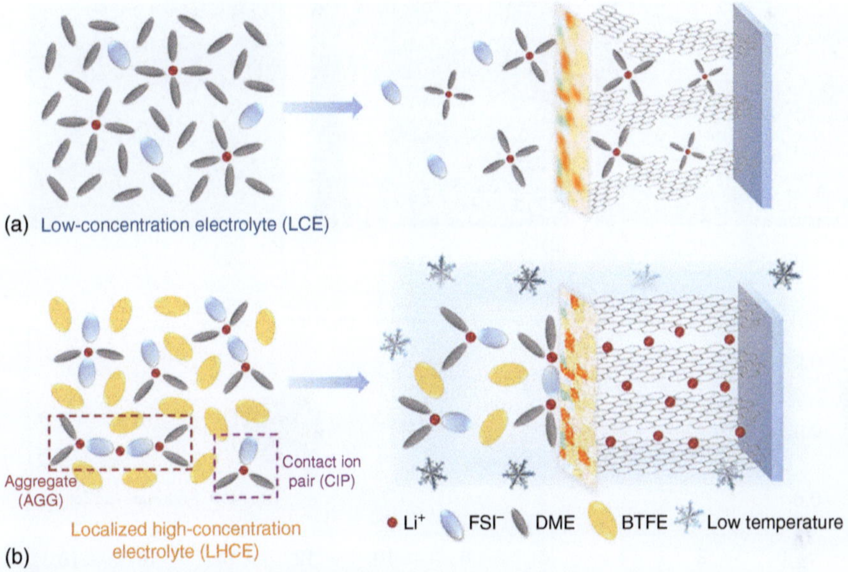

Figure 6.17 Illustrations of the solution structures and the process of Li^+ intercalation graphite layer in LCE (a) and LHCE (b) based on LiFSI in DME without/with BTFE. Source: Reproduced with permission Jiang et al. [91].

Zhang and his group added BTFE to the LiTFSI/DME electrolyte to form a stable SEI on the graphite negative electrode [91]. As shown in Figure 6.17a, in the low-concentration ether electrolyte (LCE), there are many free solvent molecules in the solvent, which are intercalated together with the Li ions into the graphite interlayer, resulting in graphite exfoliation and capacity degradation [92]. In the LHCE (Figure 6.17b), due to its special solvation structure, free solvent molecules disappear, forming contact CIPs and AGGs [93]. In addition, anions preferentially decompose on the graphite surface to form a uniform and stable SEI film, which can significantly inhibit the cointercalation behavior of ether solvents, thereby achieving excellent cycle stability of the graphite anode [91].

In the work of Cao et al. tris(2,2,2-trifluoroethyl)orthoformate (TFEO) was proposed as a diluent, which can minimize Li consumption and pulverization in LMB [54]. As shown in Figure 6.18a–c, after 100 cycles, the SEM image of the cross section of the Li metal electrode shows three different situations. In the 1 M LiFSI/DME-TFEO electrolyte, the Li metal electrode can still retain about 47 μm of native Li, while in the other two electrolytes (1 M $LiPF_6$/EC-EMC + 2%VC, 1 M LiFSI/DME), the electrodes are almost completely reacted and the volume expansion is significant. In addition, the capacity of the Li metal electrode of 1 M LiFSI/DME-TFEO electrolyte after 100 cycles was compared with the fresh Li metal electrode (Figure 6.18d), and it was found that the lost capacity was consistent with the result of Figure 6.18c. Figure 6.18e is a schematic diagram of the expansion of the Li metal electrode. When it is not cycled, the thickness of the Li metal electrode is 50 μm. After the cycle, the Li metal in the traditional electrolyte is almost

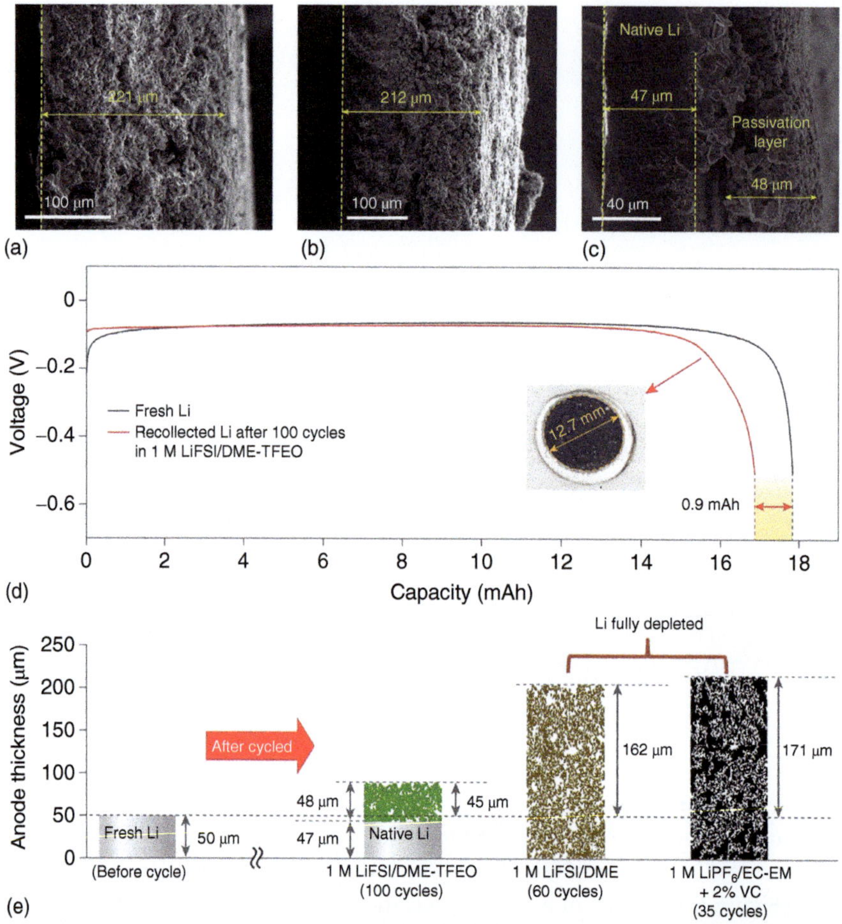

Figure 6.18 Cross-sectional views of SEM images of Li metal electrodes cycled with electrolytes of 1 M $LiPF_6$/EC-EMC + 2% VC (a), 1 M LiFSI/DME (b), and 1 M LiFSI/DME-TFEO (c); (d) quantification of Li loss during 100 cycles in 1 M LiFSI/DME-TFEO. (e) Schematic of Li loss and corresponding thickness (volumetric) expansion after 100 cycles in Li||NMC811. Source: Cao et al. [54]/with permission of Springer Nature.

completely consumed, and the thickness expands significantly. In the LHCE (1 M LiFSI/DME-TFEO), a dense and uniform SEI can be formed, which can prevent the continuous side reaction of Li and the electrolyte, thereby effectively inhibiting the volume expansion of the Li metal electrode. This is beneficial to improve the cyclic stability of LMBs and increase their safety. Subsequently, Cao et al. optimized the work of the TFEO-based electrolyte and explained the basic mechanism of the electrolyte to improve the stability of the LMA and cathode [94]. Figure 6.19a shows that as the LiTFSI concentration increases, the coordination between Li^+ and DME increases. When TFEO is further added, TFEO hardly coordinates with LiTFSI. In addition, the Raman spectra (Figure 6.19b) show that there are no free DME molecules in the LHCE, and TFEO does not change the solvation structure of the

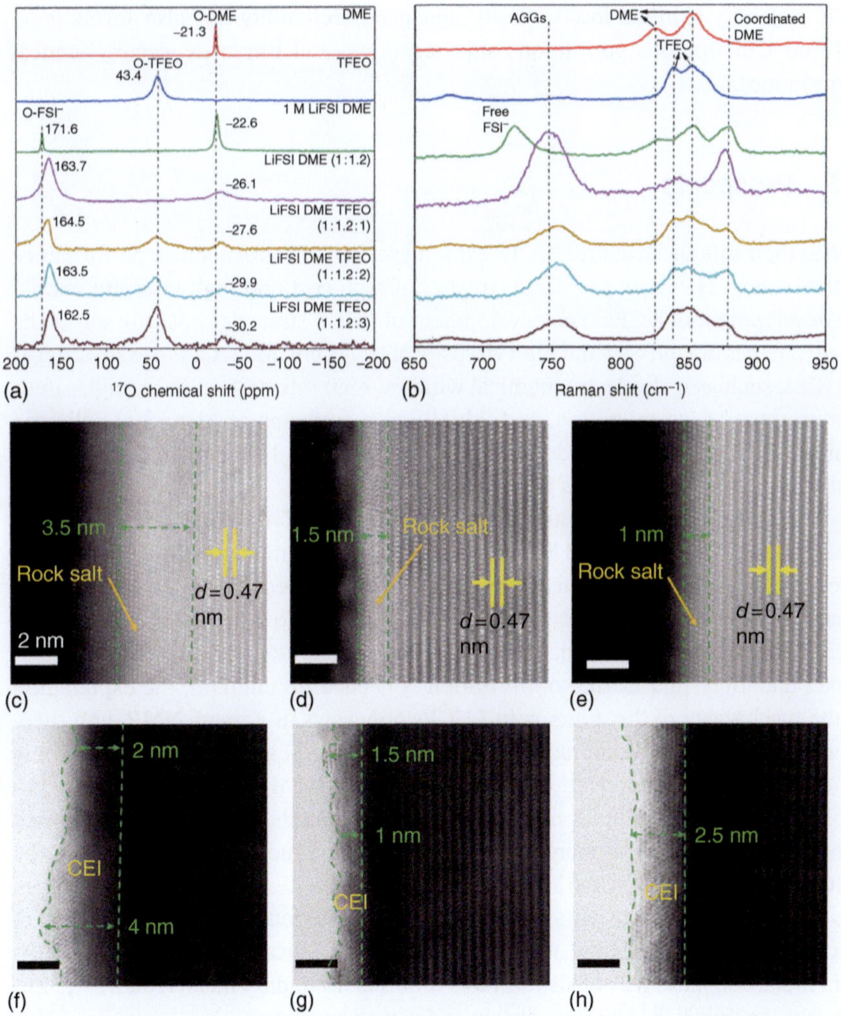

Figure 6.19 ^{17}O-NMR spectra (a) and Raman spectra (b) of different electrolytes; HAADF-STEM (c–e) and ABF-STEM (f–h) images of structural and CEI compositions of the cycled NMC811 electrodes. (c, f) LiFSI-1.2DME, (d, g) LiFSI-1.2DME-TFEO, and (e, h) LiFSI-1.2DME-2TFEO. Source: Cao et al. [94]/with permission of Elsevier.

electrolyte. The high-angle annular dark-field (HAADF)-STEM (Figure 6.19c–e) and annular bright-field scanning transmission electron microscopy (ABF-STEM) (Figure 6.19f–h) images show that a serious phase transition has occurred on the surface of the cathode material in the traditional electrolyte. The thickness of the disordered rock salt phase is 3.5 nm, and the CEI with uneven thickness is formed on the surface. In contrast, the disordered rock salt phase on the surface of the cathode material is reduced in the LHCE, and a thin and dense CEI is formed, which effectively inhibits the side reaction of the electrode and electrolyte. These results indicate that TFEO, as an effective diluent for forming the LHCE in the

HCE, not only exhibits low viscosity and good wettability but also forms good SEI and CEI, inhibits continuous side reactions, and improves electrochemical performance.

6.3 Prospects

Due to their unique structure, HCEs exhibit excellent electrochemical performance in the battery. However, the disadvantages of high cost and high viscosity restrict the development of HCEs. The development of LHCE strategies not only solves the disadvantages of high cost and high viscosity of HCEs but also retains the advantages of HCEs, such as wide electrochemical window, high rate performance, high safety, high number of ion migration, and inhibition of corrosion of Al current collector, inhibition of the growth of Li dendrites, formation of SEI. However, in LHCEs, there is still a lot of work that needs to be explored urgently.

At present, the types of diluents in LHCEs are relatively single, which cannot meet the electrolytes required by different battery application environments. Therefore, it is still necessary to find more LHCE diluents, especially LHCE diluents that meet high-temperature requirements. Although the viscosity of LHCEs is reduced, there is no relevant systematic research on the wettability of various cathode and anode materials, and further optimization is needed. In addition, the explanation of the mechanism of the diluents in LHCEs only stays in Raman, NMR, and theoretical calculations. The characterization of excellent SEI and CEI in LHCEs is only through ex situ characterization techniques such as SEM/XPS. For more in-depth exploration, there is still lack of characterization methods for in situ testing, which is not conducive to our understanding of the detailed mechanism of action in the LHCE reaction process. All in all, the development of HCEs and LHCEs has changed the traditional electrolyte pattern and opened up a new route for the development of electrolytes. At the same time, the development and application of HCEs and LHCEs in Li metal/ion batteries are also suitable for other secondary metal (Na, K, Mg, Zn) batteries, so they have better prospects and a good market.

References

1 Xu, K. (2014). *Chemical Reviews* 114: 11503–11618.
2 Zhang, J.-G., Xu, W., Xiao, J. et al. (2020). *Chemical Reviews* 120: 13312–13348.
3 Li, M., Lu, J., Chen, Z., and Amine, K. (2018). *Advanced Materials* 30: e1800561.
4 Pinson, M.B. and Bazant, M.Z. (2013). *Journal of the Electrochemical Society* 160: A243–A250.
5 Fong, R., von Sacken, U., and Dahn, J.R. (2019). *Journal of the Electrochemical Society* 137: 2009–2013.
6 Li, M., Wang, C., Chen, Z. et al. (2020). *Chemical Reviews* 120: 6783–6819.
7 Zhang, H., Eshetu, G.G., Judez, X. et al. (2018). *Angewandte Chemie International Edition* 57: 15002–15027.

8 Cheng, X.B., Zhang, R., Zhao, C.Z. et al. (2016). *Advanced Science* 3: 1500213.
9 Zhao, Q., Stalin, S., Zhao, C.-Z., and Archer, L.A. (2020). *Nature Reviews Materials* 5: 229–252.
10 Zheng, J., Engelhard, M.H., Mei, D. et al. (2017). *Nature Energy* 2: 17012.
11 Kanamura, K., Okagawa, T., and Takehara, Z.-i. (1995). *Journal of Power Sources* 57: 119–123.
12 Younesi, R., Veith, G.M., Johansson, P. et al. (2015). *Energy & Environmental Science* 8: 1905–1922.
13 Abouimrane, A., Ding, J., and Davidson, I.J. (2009). *Journal of Power Sources* 189: 693–696.
14 Chen, J., Fan, X., Li, Q. et al. (2020). *Nature Energy* 5: 386–397.
15 Su, C.-C., He, M., Shi, J. et al. (2021). *Energy & Environmental Science* 14: 3029–3034.
16 Qi, S., Wang, H., He, J. et al. (2021). *Science Bulletin* 66: 685–693.
17 Yamada, Y., Wang, J., Ko, S. et al. (2019). *Nature Energy* 4: 269–280.
18 Zheng, J., Lochala, J.A., Kwok, A. et al. (2017). *Advanced Science* 4: 1700032.
19 Jiang, G., Li, F., Wang, H. et al. (2021). *Small Structures* 2: 2000122.
20 Yamada, Y., Furukawa, K., Sodeyama, K. et al. (2014). *Journal of the American Chemical Society* 136: 5039–5046.
21 Yoshida, K., Nakamura, M., Kazue, Y. et al. (2011). *Journal of the American Chemical Society* 133: 13121–13129.
22 Nie, M., Abraham, D.P., Seo, D.M. et al. (2013). *Journal of Physical Chemistry C* 117: 25381–25389.
23 Wang, J., Yamada, Y., Sodeyama, K. et al. (2016). *Nature Communications* 7: 12032.
24 McOwen, D.W., Seo, D.M., Borodin, O. et al. (2014). *Energy & Environmental Science* 7: 416–426.
25 Yamada, Y. and Yamada, A. (2015). *Journal of the Electrochemical Society* 162: A2406–A2423.
26 Cao, X., Jia, H., Xu, W., and Zhang, J.-G. (2021). *Journal of the Electrochemical Society* 168: 010522.
27 Fan, X., Chen, L., Borodin, O. et al. (2018). *Nature Nanotechnology* 13: 715–722.
28 Jiao, S., Ren, X., Cao, R. et al. (2018). *Nature Energy* 3: 739–746.
29 Suo, L., Borodin, O., Gao, T. et al. (2015). *Science* 350: 938–943.
30 Jia, H., Xu, Y., Zhang, X. et al. (2021). *Angewandte Chemie International Edition* 60: 12999–13006.
31 Yamada, Y., Usui, K., Chiang, C.H. et al. (2014). *ACS Applied Materials & Interfaces* 6: 10892–10899.
32 Suo, L., Hu, Y.-S., Li, H. et al. (2013). *Nature Communications* 4: 1481.
33 Amine, R., Liu, J., Acznik, I. et al. (2020). *Advanced Energy Materials* 10: 2000901.
34 Cheng, X.B., Zhang, R., Zhao, C.Z., and Zhang, Q. (2017). *Chemical Reviews* 117: 10403–10473.

35 An, S.J., Li, J., Daniel, C. et al. (2016). *Carbon* 105: 52–76.

36 Zvereva, E., Caliste, D., and Pochet, P. (2017). *Carbon* 111: 789–795.

37 Ming, J., Cao, Z., Wu, Y. et al. (2019). *ACS Energy Letters* 4: 2613–2622.

38 Yao, W., Zhang, Z., Gao, J. et al. (2009). *Energy & Environmental Science* 2: 1102–1108.

39 Jeong, S.-K., Inaba, M., Iriyama, Y. et al. (2003). *Electrochemical and Solid-State Letters* 6: A13–A15.

40 Jeong, S.-K., Seo, H.-Y., Kim, D.-H. et al. (2008). *Electrochemistry Communications* 10: 635–638.

41 Nie, M.Y., Chalasani, D., Abraham, D.P. et al. (2013). *Journal of Physical Chemistry C* 117: 1257–1267.

42 Yamada, Y., Yaegashi, M., Abe, T., and Yamada, A. (2013). *Chemical Communications* 49: 11194–11196.

43 Abe, T., Fukuda, H., Iriyama, Y., and Ogumi, Z. (2004). *Journal of the Electrochemical Society* 151: A1120–A1123.

44 Goodenough, J.B. and Park, K.S. (2013). *Journal of the American Chemical Society* 135: 1167–1176.

45 Plakhotnyk, A.V., Ernst, L., and Schmutzler, R. (2005). *Journal of Fluorine Chemistry* 126: 27–31.

46 Sloop, S.E., Pugh, J.K., Wang, S. et al. (2001). *Electrochemical and Solid-State Letters* 4: A42.

47 Lux, S.F., Lucas, I.T., Pollak, E. et al. (2012). *Electrochemistry Communications* 14: 47–50.

48 Li, S., Zhang, W., Wu, Q. et al. (2020). *Angewandte Chemie International Edition* 59: 14935–14941.

49 Ye, C., Tu, W., Yin, L. et al. (2018). *Journal of Materials Chemistry A* 6: 17642–17652.

50 Liu, Q., Yang, G., Liu, S. et al. (2019). *ACS Applied Materials & Interfaces* 11: 17435–17443.

51 Sloop, S.E., Kerr, J.B., and Kinoshita, K. (2003). *Journal of Power Sources* 119–121: 330–337.

52 Liu, W., Li, J., Li, W. et al. (2020). *Nature Communications* 11: 3629.

53 Xiao, B. and Sun, X. (2018). *Advanced Energy Materials* 8: 1802057.

54 Cao, X., Ren, X., Zou, L. et al. (2019). *Nature Energy* 4: 796–805.

55 Aurbach, D., Weissman, I., Zaban, A., and Chusid, O. (1994). *Electrochimica Acta* 39: 51–71.

56 Miao, R., Yang, J., Feng, X. et al. (2014). *Journal of Power Sources* 271: 291–297.

57 Pham, T.D., Bin Faheem, A., Chun, S.Y. et al. (2021). *Advanced Energy Materials* 11: 2003520.

58 Wang, X., Yasukawa, E., and Mori, S. (2000). *Electrochimica Acta* 45: 2677–2684.

59 Morita, M., Shibata, T., Yoshimoto, N., and Ishikawa, M. (2002). *Electrochimica Acta* 47: 2787–2793.

60 Kanamura, K., Umegaki, T., Shiraishi, S. et al. (2002). *Journal of the Electrochemical Society* 149: A185–A194.

61 Yamada, Y., Chiang, C.H., Sodeyama, K. et al. (2015). *ChemElectroChem* 2: 1687–1694.
62 Peng, C.X., Yang, L., Zhang, Z.X. et al. (2008). *Electrochimica Acta* 53: 4764–4772.
63 Peng, C., Yang, L., Zhang, Z. et al. (2007). *Journal of Power Sources* 173: 510–517.
64 Kühnel, R.-S., Lübke, M., Winter, M. et al. (2012). *Journal of Power Sources* 214: 178–184.
65 Matsumoto, K., Inoue, K., Nakahara, K. et al. (2013). *Journal of Power Sources* 231: 234–238.
66 Krause, L.J., Lamanna, W., Summerfield, J. et al. (1997). *Journal of Power Sources* 68: 320–325.
67 Zhang, X.-Q., Chen, X., Cheng, X.-B. et al. (2018). *Angewandte Chemie International Edition* 57: 5301–5305.
68 Alvarado, J., Schroeder, M.A., Pollard, T.P. et al. (2019). *Energy & Environmental Science* 12: 780–794.
69 Yu, Z., Wang, H., Kong, X. et al. (2020). *Nature Energy* 5: 526–533.
70 Qian, J., Henderson, W.A., Xu, W. et al. (2015). *Nature Communications* 6: 6362.
71 Vatamanu, J., Borodin, O., and Smith, G.D. (2011). *Journal of Physical Chemistry C* 116: 1114–1121.
72 Qian, J., Adams, B.D., Zheng, J. et al. (2016). *Advanced Functional Materials* 26: 7094–7102.
73 Zhang, X., Zou, L., Xu, Y. et al. (2020). *Advanced Energy Materials* 10: 2000368.
74 Jia, H., Zou, L., Gao, P. et al. (2019). *Advanced Energy Materials* 9: 1900784.
75 Xu, G., Shangguan, X., Dong, S. et al. (2019). *Angewandte Chemie International Edition* 59: 3400–3415.
76 Jie, Y., Ren, X., Cao, R. et al. (2020). *Advanced Functional Materials* 30: 1910777.
77 Zhao, Q., Liu, X., Stalin, S. et al. (2019). *Nature Energy* 4: 365–373.
78 Chang, Z., Qiao, Y., Yang, H. et al. (2021). *Angewandte Chemie International Edition* https://doi.org/10.1002/anie.202104124.
79 Ding, J.F., Xu, R., Yao, N. et al. (2021). *Angewandte Chemie International Edition* 60: 11442–11447.
80 Moon, H., Mandai, T., Tatara, R. et al. (2015). *Journal of Physical Chemistry C* 119: 3957–3970.
81 Chen, S., Zheng, J., Mei, D. et al. (2018). *Advanced Materials* 30: e1706102.
82 Ren, X., Gao, P., Zou, L. et al. (2020). *Proceedings of the National academy of Sciences of the United States of America* 117: 28603–28613.
83 Jia, H., Gao, P., Zou, L. et al. (2020). *Chemistry of Materials* 32: 8956–8964.
84 Ren, X., Chen, S., Lee, H. et al. (2018). *Chem* 4: 1877–1892.
85 Ren, X., Zou, L., Cao, X. et al. (2019). *Joule* 3: 1662–1676.
86 Piao, N., Ji, X., Xu, H. et al. (2020). *Advanced Energy Materials* 10: 1903568.
87 Chen, S., Zheng, J., Yu, L. et al. (2018). *Joule* 2: 1548–1558.
88 Zheng, J., Chen, S., Zhao, W. et al. (2018). *ACS Energy Letters* 3: 315–321.
89 Wang, J., Yamada, Y., Sodeyama, K. et al. (2017). *Nature Energy* 3: 22–29.
90 Chen, J., Naveed, A., Nuli, Y. et al. (2020). *Energy Storage Materials* 31: 382–400.

91 Jiang, L.L., Yan, C., Yao, Y.X. et al. (2021). *Angewandte Chemie International Edition* 60: 3402–3406.
92 Wang, Z., Qi, F., Yin, L. et al. (2020). *Advanced Energy Materials* 10: 1903843.
93 Qin, L., Xiao, N., Zheng, J. et al. (2019). *Advanced Energy Materials* 9: 1902618.
94 Cao, X., Zou, L., Matthews, B.E. et al. (2021). *Energy Storage Materials* 34: 76–84.

7

Theoretical Basis for Electrolyte and Electrode Study

7.1 Redox Potential

7.1.1 Theoretical Basis

A molecule or multimolecular system all have a certain ability to gain and lose electrons. From the perspective of chemical theory, the oxidation and reduction potentials of a system can be used to compare the ability to gain and lose electrons. This is a very important part of electrochemistry. In the battery system, its essence is the mutual conversion between chemical energy and electrical energy. If the transformation process is carried out in a thermodynamically reversible manner, it is called a reversible battery. At this time, the battery is working in an equilibrium state or is infinitely close to an equilibrium state. Therefore, under isothermal and isostatic conditions, when the system changes, the reduction in Gibbs free energy of the system is equal to the maximum non-expansion work performed externally, which is expressed by the formula

$$(\Delta_r G)_{T,p} = W_{f,\max} \tag{7.1}$$

If the non-expansion work is only electrical work, Eq. (7.1) can be written as

$$(\Delta_r G)_{T,p} = -nEF \tag{7.2}$$

It is an important bridge between electrochemistry and thermodynamics. Through this formula, the absolute value of the oxidation and reduction potential can be calculated by calculating the change in the free energy of the solute before and after the gain and loss of electrons in the solvent system.

The specific process for calculating the oxidation–reduction potential is shown in Figure 7.1 [1]. In the calculation of the oxidation–reduction potential, the relationship of the difference between the free energy of the oxidation state and the reduction state of the solute in the solvent environment and the absolute value of the oxidation–reduction potential is used. $\Delta G_S^{\circ}(M^+)$ represents the free energy of solvation in the oxidation state of M, $\Delta G_S^{\circ}(M)$ represents the free energy of solvation of M, F is the Faraday constant, and ΔG_e° represents the free energy in the gas phase.

Liquid Electrolyte Chemistry for Lithium Metal Batteries: Design, Mechanisms, Strategies, First Edition. Jianmin Ma.

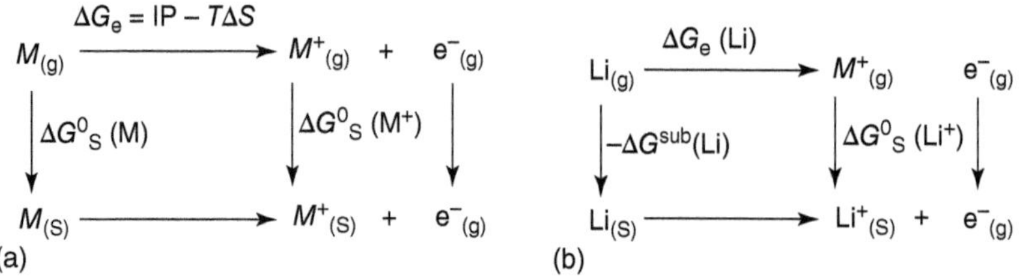

Figure 7.1 Schematic diagram of free energy solving redox potential: (a) Solving the oxidation-reduction potential by free energy; (b) Solving the oxidation-reduction potential of lithium. Source: Borodin et al. [1].

The absolute value of the oxidation–reduction potential can be calculated as follows:

$$E^{\circ}_{\text{abs}}(M) = [\Delta G_e + \Delta G^{\circ}_S(M^+) - \Delta G^{\circ}_S(M)]/F \tag{7.3}$$

In electrochemistry, to compare redox potentials, a reference electrode potential will be taken first. In experiments, the commonly accepted reference electrode potential is the hydrogen standard electrode potential. Therefore, the calculated oxidation–reduction potential is usually first subtracted from the middle value of 4.42–4.44 V [2], which is the absolute value of the hydrogen standard electrode potential obtained at room temperature. The value obtained in this way is the electrode potential relative to the hydrogen standard electrode, so that the magnitude can be compared again. In the lithium battery system, the electrode potential of Li^+/Li is usually used as the reference potential. In an aqueous solution, when Li^+/Li is used as the reference potential, 3.05 V needs to be subtracted. It can be derived from the following formula:

$$\begin{aligned} E^{\circ}\left(\text{vs}\frac{\text{Li}^+}{\text{Li}}\right) &= E^{\circ}_{\text{zbs}} + E^{\circ}_{\text{abs}}\left(\frac{\text{Li}^+}{\text{Li}}\right) = E^{\circ}_{\text{abs}} - 4.42 + 3.05\ \text{V} \\ &= E^{\circ}_{\text{abs}} - 1.37\ \text{V} \approx E^{\circ}_{\text{abs}} - 1.4\ \text{V} \end{aligned} \tag{7.4}$$

Isse et al. pointed out that 1.37 V should be replaced by 1.23 V, because different solvents will lead to different conversion factors. However, in different solvents, the difference in conversion factors is between 0.1 and 0.3 V [3].

Zhang et al. used density functional theory (DFT) calculations to obtain the absolute value of Li^+/Li oxidation–reduction potential through thermodynamic cycles. In Figure 7.1b, $\Delta G^{\text{sub}}(\text{Li})$ is heat of evaporation of lithium, $\Delta G^{\circ}_s(\text{Li}^+)$ is the free energy of solvation of lithium ions in solvents, and IP is ionization of lithium in the weather free energy. Experimentally, the free energy of Li^+ after solvation in water is −5.49 eV, and the free energy after solvation of Li^+ in dimethyl sulfoxide (DMSO) solvent is −5.75 eV [4], as reported by Eduard Westphal. Calculated by DFT, the free energy of solvation in acetonitrile is −5.25 eV, which is consistent with the value obtained by Gomer et al. Therefore, the values of $E^{\circ}_{\text{abs}}(\text{Li}^+/\text{Li})$ in Eq. (7.4) are −1.56, −1.3, and −1.79 V, respectively, when water, DMSO, and acetonitrile, respectively, are used as solvents. This is not much different from −1.4 V of the international standard solvent water.

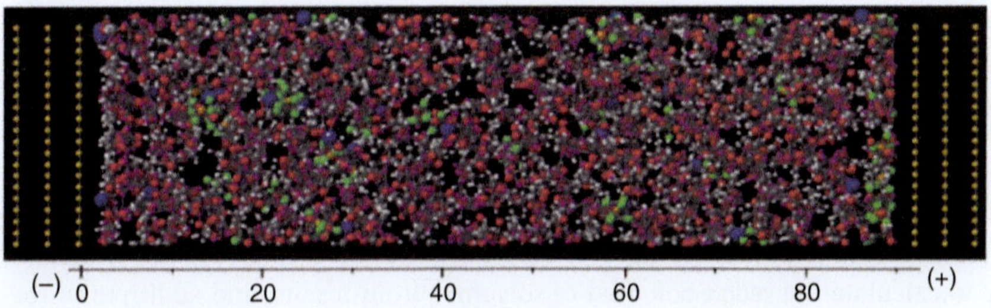

Figure 7.2 A simulated image of molecular dynamics in the electrolyte [5].

7.1.2 Solvents and Salts

In the battery system, a simulated image of molecular dynamics in the electrolyte is shown in Figure 7.2 [5]. The nature of battery charging and discharging depends on the redox reaction occurring at the two poles. When an oxidation or reduction reaction occurs, there will be a sequence for different molecules. Generally, for the molecules in the electrolyte, there are two common methods for predicting their redox sequence. One is the size of their oxidation–reduction potential. The high reduction potential is reduced first, and the low oxidation potential is oxidized first. The other one is to compare the levels of highest occupied molecular orbital (HOMO) and lowest unoccupied molecular orbital (LUMO) orbitals between different molecules.

As shown in Figure 7.3, E_g represents the electrochemical stability window of the electrolyte [6], ΦA and ΦC are the work functions of the negative and

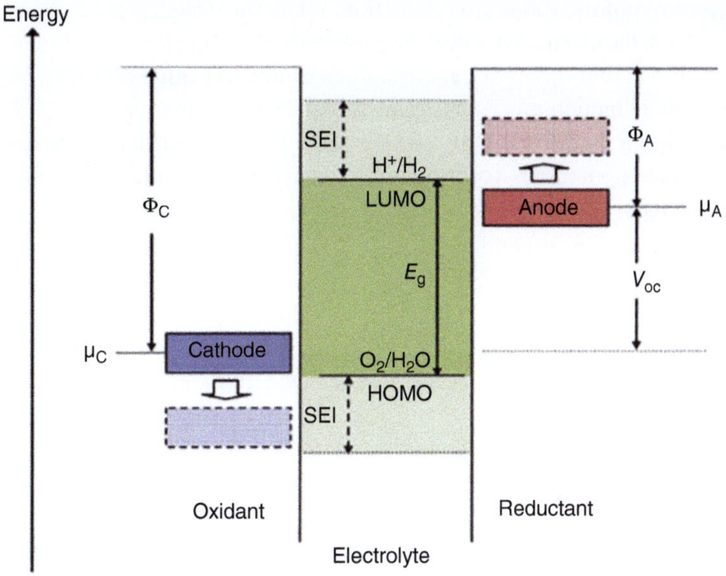

Figure 7.3 Schematic diagram of the electrochemical window and the molecular orbitals in the electrolyte. Source: Wang et al. [6]. Licensed under CC BY-4.0.

positive electrodes, and μA and μC are the oxidation–reduction potentials of the negative and positive electrodes. When the Fermi energy levels of the positive and negative electrodes are between the HOMO and LUMO energy levels of the electrolyte molecules, this is the most ideal situation: the molecules in the electrolyte do not participate in the battery's charge and discharge reaction process, only the positive and negative materials at both ends of the battery participate in the charging and discharging of the battery. Therefore, it is very important for one to calculate the redox potential of solvents, lithium salts, and additives in the electrolyte.

7.1.3 Redox Potential of the Complex

Usually in the electrolyte, there will be a solvated shell around the cations. There will be several solvent, anion, or additive molecules surrounding the lithium ion. When we study the oxidation–reduction potential of the electrolyte, only considering the oxidation–reduction potential of a single molecule may be inconsistent with real experiments. Moreover, we understand that increasing the amount of EC in the lithium ion solvation shell has no effect on the reduction potential of EC [7]. Therefore, studying the overall redox potential of the Li^+ solvation shell or the common redox potential of lithium ions and solvent molecules is closer to reality.

Figure 7.4 shows a schematic diagram of the reduction potential of a different reduction conformation of FEC [8]. Figure 7.4a,b consists of optimized schematic diagrams of the closeness of O and F on the carboxyl group of lithium ions and FEC. From the binding energy, we found that lithium ions are easier to combine with O on the carboxyl group to form a stable structure. Figure 7.4c–g is the reduction potential of different conformations after reduction. When the system gets electrons and reduction occurs, the calculated reduction potentials of lithium ions in different positions of the system and different conformations of FEC are quite different. In particular, when the reduction process is accompanied by the formation of LiF, the reduction potential of the system will increase greatly. This also further illustrates the difference in redox potentials corresponding to different conformations after reduction or oxidation.

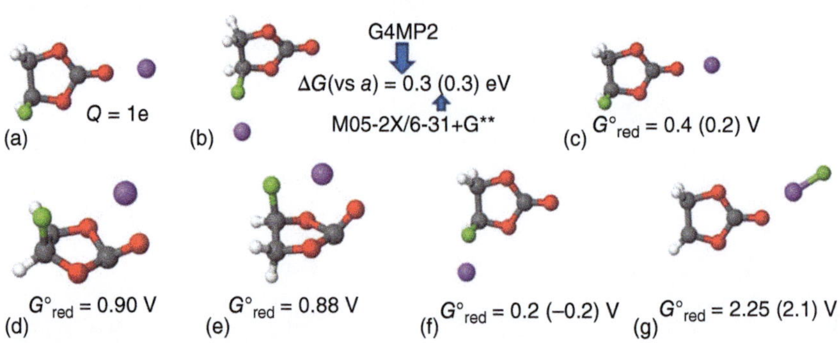

Figure 7.4 (a–g) Diagram of the reduction potential process of FEC in different reduction conformations. Source: Borodin et al. [8].

7.2 Solvation Structure

7.2.1 Basic Theory

In the battery system, the solvation structure of lithium ions has a great influence on battery performance [9]. Moreover, in addition to the battery system, the study of solvation structure has been extensively studied in the fields of chemistry and materials, and there are very interesting phenomena [10]. In particular, the difference in the solvation structure will cause the difference in the extraction and insertion of lithium ions at the lithium metal negative electrode, resulting in a great difference in the growth of solid electrolyte interphase (SEI) and cathode electrolyte interphase (CEI) [11].

To calculate the solvation structure, there are mainly two calculation methods: either using the classical force field or using the quantum method of DFT to run molecular dynamics. Both methods have their pros and cons. Classical molecular dynamics can simulate a larger system, can be less time consuming, and provide general accuracy, while quantum molecular dynamics runs more accurately, with a smaller system and can be more time consuming. Figure 7.5 a–c demonstrates the simulated images of classical molecular dynamics and quantum molecular dynamics, respectively [12, 13].

In classical molecular dynamics, the force field used for molecular simulation in electrolyte is usually the OPLS-AA force field. The force field parameters include bond parameters, angle parameters, dihedral angle parameters, and van der Waals interaction terms. The bond, angle, and dihedral potential energy forms are the simplest harmonic oscillator models. The interaction term between atoms uses the *L–J* potential. The theoretical basis of quantum molecular dynamics is still DFT, which only adds to the process of wave function changes over time, so as to simulate the system over time. Whether it is classical molecular dynamics or quantum molecular dynamics, there are many kinds of mathematical solving processes for its body on a computer. For different systems, although the methods or force fields may be different, the essential solution process is to find the change of the wave function with time or the change of the state of the system with time.

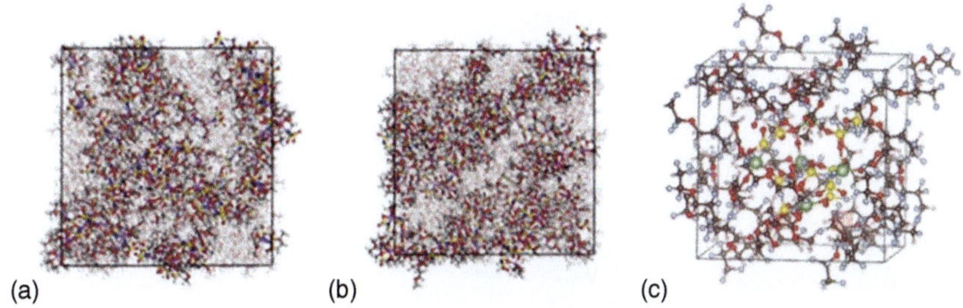

Figure 7.5 (a and b) 10 ns and 15 ns classical and (c) quantum molecular dynamics simulations. Source: Zhang et al. [12] and Zhang et al. [13].

7.2.2 Influencing Factors and Implicit Solvent Model

For an electrolyte system, when we predict the solvation structure of lithium ions, or guess whether an additive molecule will participate in the solvation structure of lithium ions, there are two main influencing factors. One is the polarity of the molecule in the solution, and the other is the binding energy of the molecule's lithium ions. When the binding energy of a molecule in the electrolyte with lithium ions is relatively large or the polarity of the molecule is relatively large, it indicates that the molecule is likely to participate in the solvation structure of lithium ions. The polarity of a molecule can be judged by the size of a dipole moment of the molecule.

Figure 7.6 is a comparison diagram of the dipole moments of solvent molecules and additive molecules in an electrolyte system [14]. From this, we can speculate that the introduction of additive molecules is likely to have a certain impact on the lithium-ion solvation shell because the dipole moment of the additive molecule is still relatively large, and it is likely to participate in the first solvent of lithium ions in the shell. When the lithium-ion solvation shell changes, when lithium ions are transported to the surface of the negative electrode, and when the electrolyte undergoes reductive decomposition, the composition of SEI may be different, which will have a great impact on the electrochemical performance of the battery.

In addition to considering the influence of molecular polarity on the solvation structure, another important factor the binding energy of lithium ions and molecules cannot be ignored. The expression of binding energy is $\Delta E = E_{ab} - E_a - E_b$, and the total energy when two molecules are in contact is subtracted from the energy when two molecules exist alone. As shown in Figure 7.7a–c, when we calculate the binding energy, we can calculate the binding energy of the molecule under vacuum, and we can also calculate the binding energy after adding the solvent model [15]. From the figure, we can find that when considering the solvent model, the binding energy will generally be smaller than the binding energy under vacuum, and the order of the magnitude of the binding energy is almost the same.

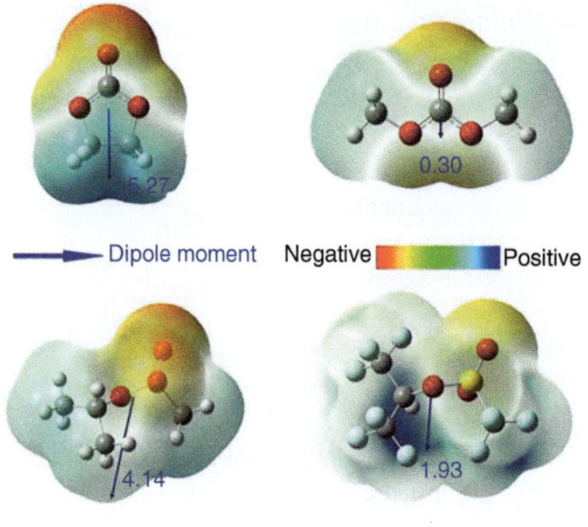

Figure 7.6 Schematic diagram of molecular dipole moment. Source: Li et al. [14].

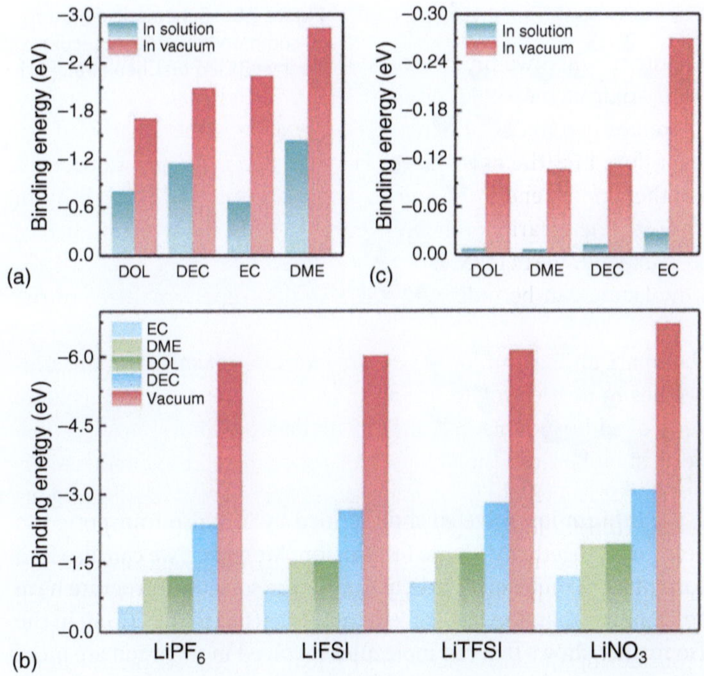

Figure 7.7 Binding energy of molecules in vacuum and solution: (a) Comparison of the binding energies of lithium ion with different solvents in vacuum and solution; (b) Comparison of the binding energies of lithium ion with different anions in vacuum and solution; (c) Binding energy between each solvent. Source: Based on Chen et al. [15].

When considering the effect of solvent on binding energy, we generally use the implicit solvent model, and the solvent model uses the SMD model. The implicit solvent model does not specifically describe the specific structure and distribution of solvent molecules near the solute, but simply considers the solvent environment as a polarizable continuous medium. The advantage of considering the solvent effect is that the average effect of the solvent can be expressed without considering various possible solvent-layer molecular arrangements like the explicit solvent model, and it will not increase the calculation time.

7.2.3 Solvation Analysis

Whether through classical molecular dynamics or quantum molecular dynamics, we all simulate the actual electrolyte system with the computer. We can use some analysis software to post-process the data of the nature of the system. In the simulated system, due to the difference in molecular polarity and binding energy with lithium ions, some molecules will enter the solvation shell of lithium ions or become free solvent molecules. Some scientific researchers believe that the SEI film formed on the surface of the lithium negative electrode has a great relationship with the molecules involved in the solvation when lithium ions are inserted and extracted from the lithium negative electrode. They believe that in the process of lithium-ion transport

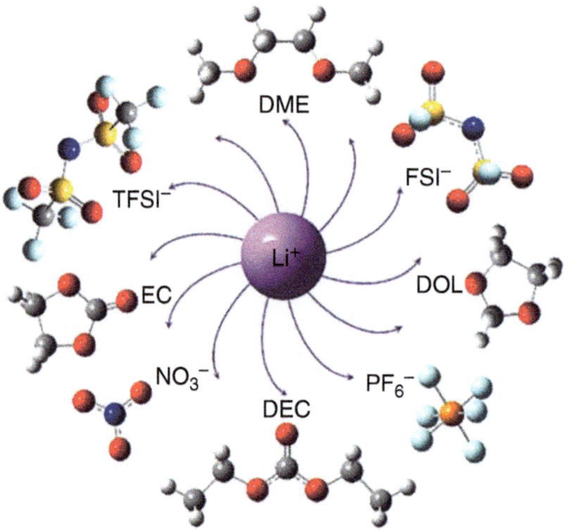

Figure 7.8 Schematic diagram of common solvated molecules. Source: Based on Chen et al. [15].

in the electrolyte, the lithium-ion solvated shell formed by it is also transported to the surface of the electrode together with the lithium ion. Moreover, we can also find through calculations that the molecules participating in the solvation structure have a higher reduction potential than the free solvent molecules not participating in the solvation. This also further shows that the molecules involved in solvation are more likely to be reduced and thus participate in the formation of the SEI film. However, because the chemical reactions that occur in the electrolyte are very complicated, it cannot be simply assumed that free molecules will not participate in the formation of SEI. The general opinion is that the probability of free molecules participating in film formation is very small (Figure 7.8).

When we deal with molecular dynamics data, one of the most important properties is the statistical distribution of the radial distribution function of the trajectory structure. Figure 7.9 a–c shows the graphs of the most common radial distribution function [16]. The radial distribution function represents an image of the probability of occurrence of another specific atom around a certain atom. Therefore, we can represent one of the atoms as lithium ions, and the other atom is the atom with the most negative charge of a molecule in the electrolyte to calculate the radial distribution function of these two atoms and get the molecular distance from lithium. Electrolyte molecules are distributed around ions, forming solvation. The radial distribution function of the molecule and the lithium ion is used to analyze whether the molecule participates in the solvation and the average degree of coordination. Therefore, through the radial distribution function of molecules and lithium ions, we have an intuitive understanding of the solvation of lithium ions in the electrolyte and the distribution of various molecules in the electrolyte.

7.2.4 De-Solvation

Molecular dynamics methods have also been used to study the de-solvation behavior at the SEI interface. Jorn et al. [17] conducted MD simulations on the surface of graphite electrodes with or without SEI, and showed that under the action of an electric field, Li^+ will be closer to SEI as the thickness of LiF and SEI increases.

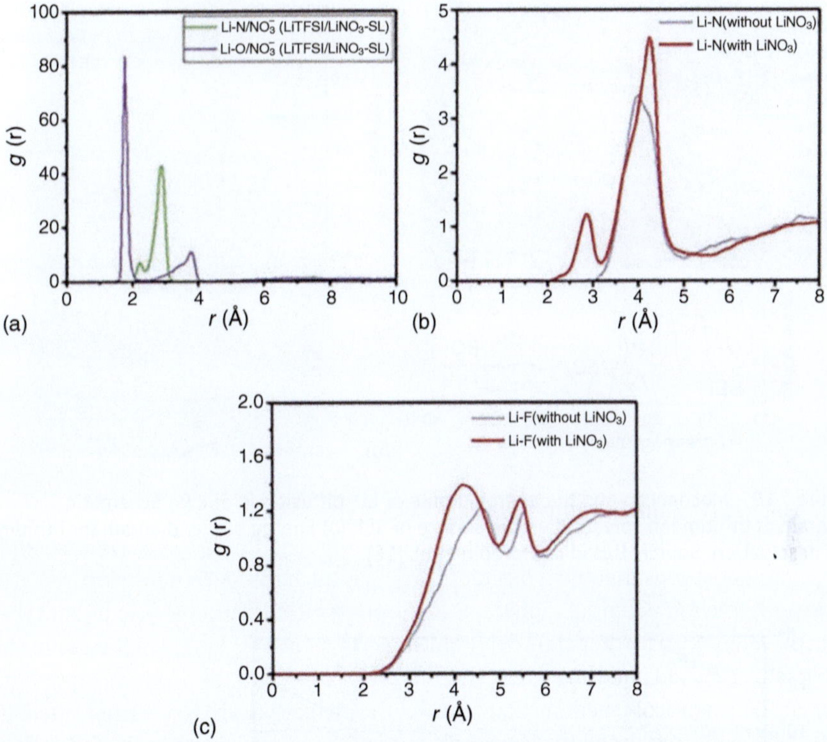

Figure 7.9 (a) Li-N (nitrate anion) and Li-O graph; (b) Li-N (TFSI anion) and (c) Li-F (TFSI anion) graphs of radial distribution function. Source: Fu et al. [16].

As shown in Figure 7.10 a and b, Borodin and Bedrov [18] subsequently performed MD simulations on Li_2EDC and Li_2BDC and predicted that the de-solvation energies of Li^+ at the Li_2EDC/electrolyte and Li_2BDC/electrolyte interfaces were 0.42 and 0.46 eV, respectively.

However, due to the complexity of SEI component and structure, precisely modeling the de-solvation process in SEI is far not reachable for molecular dynamics. The simulation of single component only gives a rough description of the de-solvation process, even the most state-of-the-art in situ characterization method could not give a quantitative description of SEI. The homology modeling of molecular dynamics also could not give the information of SEI properties without quantitative description.

Another indirect way to model SEI is to use ab initio MD to simulate decomposition of the electrolyte near the electrode. Budzien and Leung [19] conduct ab initio MD to trace the initial EC decomposition process on graphite, and confirmed two pathways of EC reduction. However, due to the limitation of the AIMD scale, the subsequent interaction among reduction products could not be observed.

Favorably, deep learning provides an opportunity to enlarge the scale of molecular dynamics without affecting the precision. Recently, Zhang et al. [20] proposed deep potential molecular dynamics (DPMD) which is based on deep neural networks to reduce the computational cost of MD (Figure 7.11).

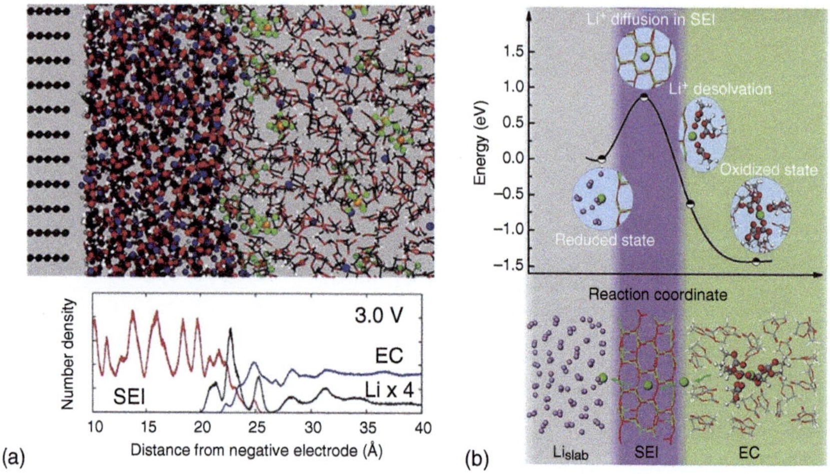

Figure 7.10 Mechanism and the energy profile of Li^+ diffusion in SEI: (a) Schematic diagram of lithium ion solvation on the surface of SEI; (b) Energy barrier diagram for lithium ion desolvation. Source: Based on Borodin et al. [18].

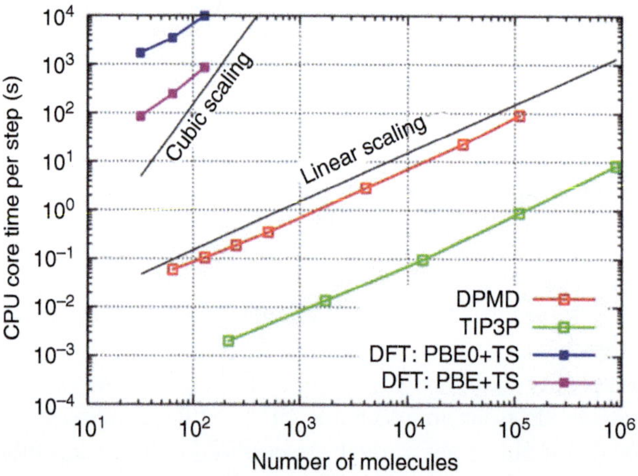

Figure 7.11 Computational cost of DPMD, empirical model, and DFT. Source: Zhang et al. [20].

7.3 Lithium Diffusion

7.3.1 Lithium Diffusion in SEI

According to experiments probing SEI components, Li_2CO_3, LiF, Li_2O, Li_2EDC, and Li_2BDC are the main components of SEI. Using DFT to modeling diffusion mechanism is important.

For Li_2CO_3, Li_2O, and LiF, Chen et al. [21] used the nudged elastic band method to investigate the Li^+ diffusion mechanism. According to DFT calculation, all

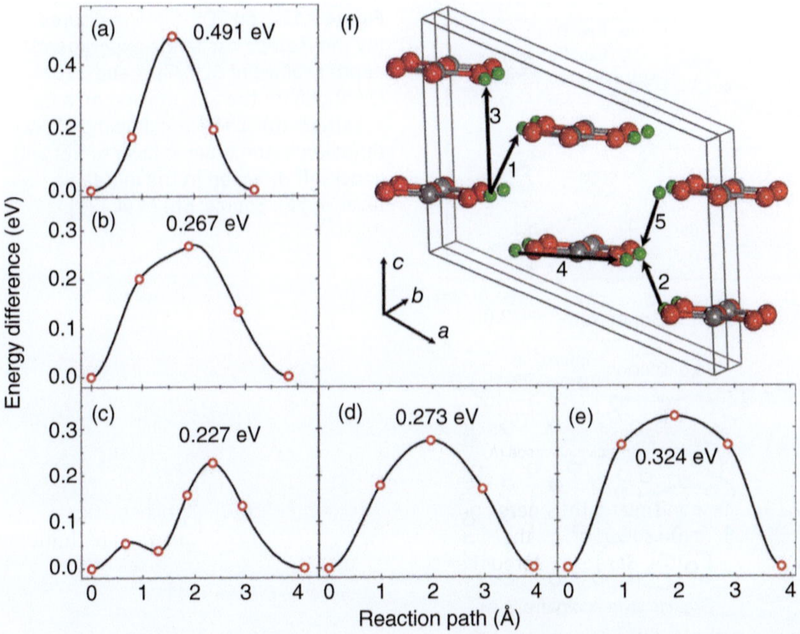

Figure 7.12 Energy profiles and migration pathways of single lithium migration in Li_2CO_3: (a–e) diffusion barriers of path 1-5; (f) schematic diagram of the diffusion of different paths. Source: Chen et al. [21].

three components are insulators; Li^+ could only transfer through the localized holes created by lithium vacancies, which do not contribute much to electronic conduction. When lithium vacancies are available, Li^+ diffusion in Li_2CO_3 and Li_2O could be fast with 0.227–0.491 and 0.152 eV energy barriers, respectively. However, Li^+ in LiF could not migrate easily due to the higher energy barrier of 0.729 eV (Figure 7.12a–f).

To be more specific and precise, Harris and coworkers [22] reported that in the inner layer of SEI, mainly Li_2CO_3 in that case, Li^+ diffuses through knock-off mechanism, which is due to the stable coordination environment rather than direct hopping through the empty spaces between the lattice sites (Figure 7.13).

Based on this mechanism, the two-layer/two-mechanism model, which indicates Li^+ diffuses in the pore of the outer organic layer and diffuses through the inner inorganic layer via knock-off mechanism, offers an applicable method to model the conductivity behavior in complex SEI films on the electrode surface.

For Li_2EDC and Li_2BDC, which are mainly the reduction products of EC, knock-off mechanism could also be proved by hybrid MD–Monte Carlo (MC) scheme with a many-body polarizable force field (APPLE&P) [23]. Both ordered and disordered Li_2EDC/Li_2BDC show a similar Li^+ migration energy barrier of about 0.6 eV when fitted to the Arrhenius relationship below 500 K. When temperature is lower than 450 K, both of these materials are glassy Li^+ de-solvated with cations.

LiF and Li_2O also play a crucial role in inner SEI. As discussed before, migration energy barrier of Li^+ in LiF is higher than that of Li_2CO_3 and Li_2O. Pan et al.

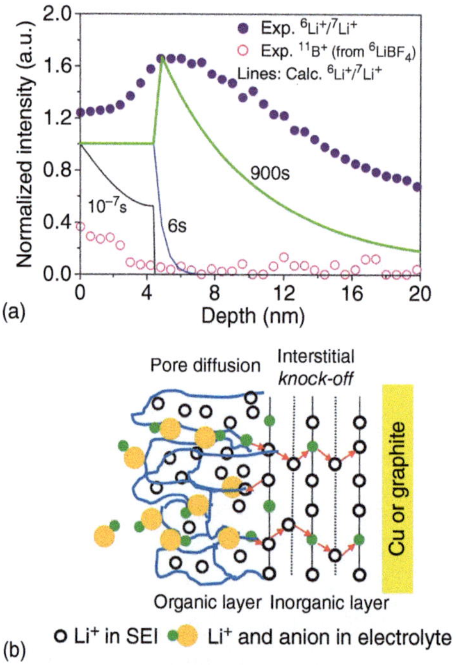

Figure 7.13 (a) TOF-SIMS measured (by the isotope exchange experiment) depth profiles of $6Li^+/7Li^+$ and $11B^+$ (symbols) for the SEI growing on a Cu substrate. (b) Schematic drawing of pore diffusion in the organic layer of SEI and knock-off diffusion in the inorganic layer of SEI. Source: Shi et al. [22].

[24] developed a method to investigate defects in LiF. The defect formation in SEI is determined by an open-circuit voltage of the electrode and Li^+ diffuses through Li^+ vacancy hopping rather knock-off mechanism.

7.3.2 Lithium Diffusion in Electrode Material

7.3.2.1 Calculation of Electrode Materials

Driven by the increasing demand for electrochemical energy storage, the development of high-energy-density lithium-ion batteries has been the direction of scientists' efforts for decades. However, the lack of high-energy-density electrode materials has become a key problem to be solved. Computational simulation plays an important role in finding new functional materials. DFT can be used to predict the performance of materials in battery applications. DFT solves the Schrödinger equation approximately, which was first developed in 1964 by Hohenberg and Kohn. DFT can not only help to explain the characteristics of electrode materials but also play an active role in predicting new electrode materials with different characteristics. The growth of computing power will enable millions of materials to be screened for battery applications, find suitable electrode materials in advance, reduce the failure of experiments, and explain the experimental results in theory.

7.3.2.2 Equilibrium Voltage

One of the important characteristics of electrode materials is expected voltage, which determines the energy stored per unit charge. Lithium-ion intercalation voltage is an important parameter of lithium-ion batteries, and the ideal electrode material is that the voltage platform should be sufficiently high for the cathode material and

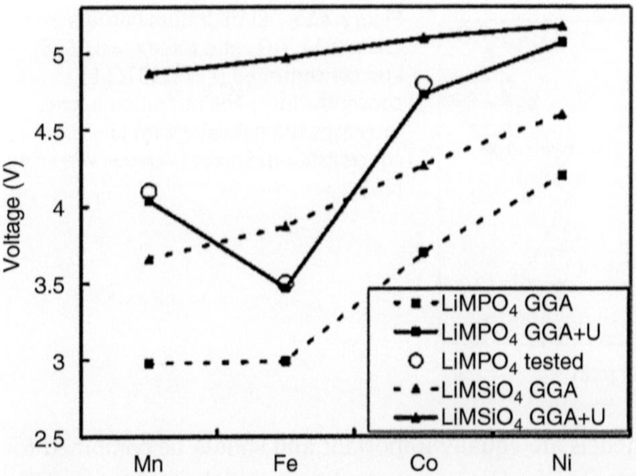

Figure 7.14 Voltage for M = Mn, Fe, Co, and Ni in olivine phosphates $LiMPO_4$ (squares) and silicates $LiMSiO_4$ (triangles), as predicted by GGA (dotted line) and GGA + U (solid line). Available experimental values for $LiMPO_4$ (M = Mn, Fe, Co) are also shown as big circles. Source: Zhou et al. [25].

low for the anode material, to get higher working voltage, and the ideal material should have higher energy density for lithium-ion batteries. In the first principle, the average intercalation voltage (AIV) can be calculated by calculating the total electron energy of the ground state of the material, which is close to the experimental voltage. As shown in Figure 7.14, the calculated voltage of $LiNiPO_4$ cathode material is 5.1 V by Zhou et al., while the experimental value is 5.1–5.3 V [25].

7.3.2.3 Ionic Mobility

Rate performance refers to the current value of the battery discharging its rated power in a certain period. The higher the rate performance of the battery, the shorter the discharge time of the same capacity, which is conducive to the rapid charge and discharge of the battery. By using DFT, we can calculate the diffusion energy barrier of lithium ions in materials by using the NEB (nudged elastic band) method, and the diffusion energy barrier corresponds to the diffusion energy of lithium ions, that is, the diffusion rate. The lower the diffusion barrier, the higher the diffusion rate and the rate performance is higher. As shown in Figure 7.15, Ven et al. [26] calculated the diffusion barrier of lithium ions in $LiTiO_2$, which shows that ion conductivity strongly depends on the concentration of Li.

7.4 Conclusion

Multiscale modeling of electrolytes, especially the three main aspects discussed in this chapter, contributed qualitative and quantitative insights to understand the effects of electrolytes in the battery system. We reviewed current modeling efforts based on ab initio and empirical theory from the perspectives of prediction. Both

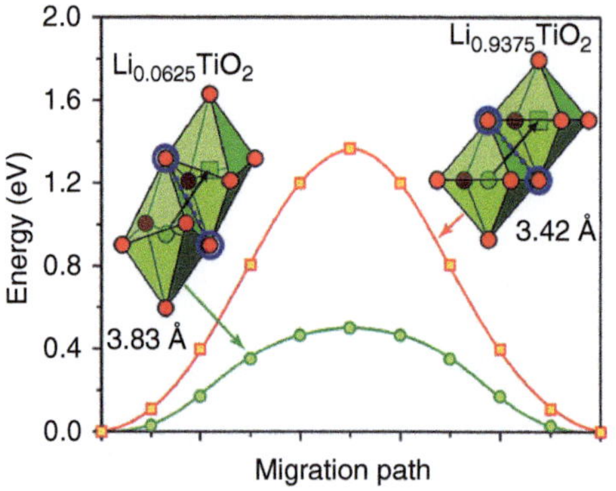

Figure 7.15 Li migration barriers in anatase Li_xTiO_2 at dilute ($x = 0.0625$) and concentrated ($x = 0.9375$) Li concentrations. The migration barrier increases dramatically with Li concentration. Source: Van der Ven et al. [26].

ab initio and empirical models are equally important and should be combined for qualitative and precise prediction. However, simplified models could not cover the whole complex interaction in the battery system; a proper combination of multiscale models and new approaches to discover complex systems is being simultaneously developed. Modeling the multicomponent and multifactor influence in the battery system is still a great challenge for the research community. Discovery of new battery systems depends on diligent research by experimental scientists and theoretical scientists in various areas.

References

1 Borodin, O., Behl, W., and Jow, T.R. (2013). *The Journal of Physical Chemistry C* 117 (17): 8661–8682.

2 Trasatti, S. (1986). *Pure Appl. Chem.* 58 (7): 955–966.

3 Isse, A.A. and Gennaro, A. (2010). *J. Phys. Chem. B* 114 (23): 7894–7899.

4 Westphal, E. and Pliego, J.R. (2005). *J. Chem. Phys.* 123 (7): 074508.

5 Vatamanu, J., Borodin, O., and Smith, G.D. (2012). *The Journal of Physical Chemistry C* 116 (1): 1114–1121.

6 Wang, A., Kadam, S., Li, H. et al. (2018). *npj Computational Materials* 4 (1): 15.

7 Cresce, A.v.W., Borodin, O., and Xu, K. (2012). *J. Phys. Chem. C* 116 (50): 26111–26117.

8 Borodin, O., Olguin, M., Spear, C.E. et al. (2015). *Nanotechnology* 26 (35): 354003.

9 Xing, L., Zheng, X., Schroeder, M. et al. (2018). *Acc Chem. Res.* 51 (2): 282–289.

10 Israelachvili, J. and Wennerstrom, H. (1996). *Nature* 379 (6562): 219–225.

11 Wang, L., Zhou, Z., Yan, X. et al. (2018). *Energy Storage Materials* 14: 22–48.

12 Zhang, X.-Q., Chen, X., Hou, L.-P. et al. (2019). *ACS Energy Letters* 4 (2): 411–416.

13 Zhang, X., Zou, L., Xu, Y. et al. (2020). *Advanced Energy Materials* 10 (22): 2000368.

14 Li, X., Liu, J., He, J. et al. (2021). *Advanced Functional Materials* 30 (37): 2104395.

15 Chen, X., Zhang, X.-Q., Li, H.-R., and Zhang, Q. (2019). *Batteries & Supercaps* 2 (2): 114–114.

16 Fu, J., Ji, X., Chen, J. et al. (2020). *Angewandte Chemie International Edition* 59 (49): 22194–22201.

17 Jorn, R., Kumar, R., Abraham, D.P., and Voth, G.A. (2013). *The Journal of Physical Chemistry C* 117 (8): 3747–3761.

18 Borodin, O. and Bedrov, D. (2014). *The Journal of Physical Chemistry C* 118 (32): 18362–18371.

19 Leung, K. and Budzien, J.L. (2010). *Physical Chemistry Chemical Physics* 12 (25): 6583–6586.

20 Zhang, L., Han, J., Wang, H., and Car, R.; E, W (2018). *Physical Review Letters* 120 (14): 143001.

21 Chen, Y.C., Ouyang, C.Y., Song, L.J., and Sun, Z.L. (2011). *The Journal of Physical Chemistry C* 115 (14): 7044–7049.

22 Shi, S., Lu, P., Liu, Z. et al. (2012). *Journal of the American Chemical Society* 134 (37): 15476–15487.

23 Bedrov, D., Borodin, O., and Hooper, J.B. (2017). *The Journal of Physical Chemistry C* 121 (30): 16098–16109.

24 Pan, J., Cheng, Y.-T., and Qi, Y. (2015). *Physical Review B* 91 (13): 134116.

25 Zhou, F., Cococcioni, M., Kang, K., and Ceder, G. (2004). *Electrochemistry Communications* 6 (11): 1144–1148.

26 Van der Ven, A., Bhattacharya, J., and Belak, A.A. (2013). *Accounts of Chemical Research* 46 (5): 1216–1225.

8 Outlook

Lithium metal with highest theoretical capacity is considered the "gold" anode of rechargeable Li batteries. However, Li metal anodes suffer severe limitations due to irregular Li dendrite growth, unstable solid electrolyte interphase (SEI) formation, low CE, and poor cycle life. In particular, irregular formation of Li ions is still the biggest obstacle to the commercialization of Li metal anodes. In the recent decades, researchers have proposed a series of dendrite growth models and experiments to describe and study the evolution behavior of polymorphic Li, which has been used to explore effective routes to delay/suppress/redirect the ordered Li dendritic growth. Electrolyte modifications (such as electrolyte additive selection, safe electrolytes, high-pressure electrolytes, extreme temperature electrolytes, high-concentration electrolytes) play a significant role in stabilizing Li metal anodes. Therefore, it is of great significance to optimize these strategies for efficient and safe Li metal anodes and integrate them into a practical battery system.

Electrolytes have a significant impact on properties of rechargeable Li batteries. However, due to the harsh environment under industry-demanding levels, evolution of Li metal is much more complicated and difficult to predict and control; there is still a long way to go forward for both fundamental study and future industrial commercialization. Here, we provide our outlook with respect to Li metal anodes for constructing safe and better Li metal batteries (LMBs).

(1) Uncontrolled dendrite growth and poor cycle efficiency are the biggest obstacles to the practical application of LMBs. Fortunately, the inhibition of dendrite formation and the growth of nondendrites have been achieved in many systems, and high coulombic efficiency of Li||Cu cells (>99.5%) has also been achieved in ether-based electrolytes. However, the lack of high voltage resistance of ether electrolytes severely limits the improvement of energy density of batteries. Therefore, on this basis, the development of high-voltage-resistant ether electrolytes is a key to breakthrough high-energy-density Li metal batteries.

(2) Due to high reactivity of Li metal anodes and the complex structural properties of the Li metal interface, ex situ characterization can sometimes degrade the original performance and lead to some misleading results. Regarding the behavior of Li ion deposition and the evolution of Li deposition morphology

Liquid Electrolyte Chemistry for Lithium Metal Batteries: Design, Mechanisms, Strategies, First Edition. Jianmin Ma.

(including whiskers, moss, trees, or mixed) and the chemical properties of SEI, more advanced nondestructive in situ characterization techniques (such as X-ray tomography, synchrotron radiation, cryo-electron microscopy, in situ TEM) are required to track, which help to understand the nucleation and growth mechanism of dendrites at a deeper level, while it can also realize real-time dynamic and accurate detection of SEI changes in operation, helping unearth the puzzle of SEI. In addition, we should encourage more cross-disciplinary exchanges and cooperation, so that academia and industry can learn from each other more and better. Only by recognizing the most basic mechanism, we can further design a better SEI and achieve high-energy-density rechargeable Li metal batteries for large-scale applications.

(3) In the process of repeated deposition and peeling of Li metal batteries, dead Li is formed between the SEI and electrically isolated metallic Li, causing a large amount of Li loss. The continuous accumulation of dead Li increases impedance, which will lead to heat loss, while Li loss and dead Li debris are the primary causes of performance degradation in Li metal batteries. It will be a very important research topic to develop more efficient strategies to regenerate dead Li and restore activity to compensate for the loss of Li, thereby improving the overall performance of Li metal batteries.

(4) The formation mechanism of SEI and the diffusion mechanism of Li ions in SEI still need to be further studied, which will be of great help to researchers to design excellent batteries. Recently, the boom in robot learning and artificial intelligence has attracted widespread attention in the research community, and whether it can produce unexpected results if applied to SEI research needs to be studied. If the structure of the SEI can be revealed at the atomic level through ultrahigh-resolution characterization techniques such as scanning tunneling microscopy (STM), it could promote the birth of safer and more high-energy-density batteries in the future.

(5) The structure and components of SEI in Li metal anode have always been a very important topic. However, many issues about the components and spatial distribution of SEI should be further systematically studied to determine what species are formed, how they form on the anode surfaces, and how to protect Li metal anodes. In addition, the effect of LiF in the SEI is still ambiguous and needs a huge amount of experimental data to support it. Thus, the true structure of the SEI remains a big challenge.

(6) On the basis of the above research, most of the work uses the coin cell to evaluate the performance of Li metal batteries, and it is rarely considered to be researched in high-energy-density pouch or cylindrical cells. More efforts should be dedicated to use pouch or cylindrical cells to evaluate the stability of Li metal anodes, which are closer to the actual situation. In addition, to save production costs and improve the energy density of the cells, the amount of electrolyte used in practical commercial batteries is much less than that used in the laboratory research. Insufficient electrolytes will easily lead to concentration polarization, which further results in the interface reaction and

uncontrollable mass transport or irregular growth of Li dendrites. Therefore, the effect of electrolyte amount in cells should be given more attention in the future.

(7) Mostly, organic liquid electrolytes are used in Li metal batteries, which have the risk of leakage and flammability and explosion, which can cause serious safety problems. In addition, instability of the interface between the highly active Li metal and organic liquid electrolytes often leads to poor chemical stability. Solid electrolytes are considered to eliminate these hidden dangers due to organic liquid electrolyte safety issues, and they show high mechanical properties to solve compatibility issues related to Li metal anodes. However, low ionic conductivity and unstable interface with Li metal limit their practical application in Li metal batteries. Therefore, more efforts should be invested on solid electrolytes to make breakthroughs in future research.

(8) In most research works, an extremely thick Li foil was used as the counter electrode of Li metal batteries, and the amount of Li used in the actual cycle was less than one-tenth, which means that the Li content is seriously excessive. This cannot be used in an actual battery system. Therefore, anode-free full cell systems are strongly recommended to evaluate the whole performance of Li metal plating/stripping. Generally, without using air- and moisture-sensitive Li metal, anode-free cells are superior to traditional Li metal batteries in terms of energy density, simplified structure, and minimal cost. However, poor cycle life, low coulombic efficiency, uncontrolled metal dendrite growth, and unstable SEI severely limit the practical application of anode-free batteries. Therefore, most strategies for stabilizing Li metal anodes can be extended to anode-free systems.

(9) It is well known that high-activity Li metal will react when it comes in contact with air, which also has high requirements for the characterization of Li anodes. Therefore, LiX alloys (X = Mg, Sn, or Ag) that are stable in air can be considered for Li metal anodes. LiX alloys have long-term air stability, which brings great convenience in characterization. So far, some breakthroughs have been made by converting air-sensitive Li metal into LiX alloys, but there is still a huge research space.

(10) In recent years, fires and explosions are caused frequently by batteries , which shows that Li batteries still have very serious safety problems. Therefore, safe Li metal batteries are the ultimate standard for practical applications. In this regard, there have been many studies on flame-retardant electrolyte systems in Li metal batteries. However, electrochemical stability of flame-retardant electrolytes in cathode and anode electrodes is severely restricted. Therefore, to build a safe rechargeable battery system, more efforts should be devoted to achieve high efficiency, high flame retardancy, and excellent electrochemical stability.

(11) Advancements in battery technology are largely driven by the discovery of new materials – the discovery of $LiCoO_2$ has totally changed the world. In the information age, we have the opportunity to discover new materials in a new paradigm. Today, materials discovery involves a more number of trial

and error; it usually takes decades to apply a new material in the battery system. The long period is basically due to the complexity of multidimensional optimization problems. In addition, most of the data produced in this procedure are getting wasted. We need a scalable approach that could leverage the talent and efforts of the whole materials community. There are many theories that could predict important properties of materials using quantum chemical approximations such as density functional theory (DFT). Virtual testing of materials could efficiently help optimize and design of materials on the computer, which could significantly shorten the long discovery process. To support high-throughput computational approaches, huge effort has been invested in establishing a materials database.

(12) For inorganic crystal materials, Inorganic Crystal Structure Database (ICSD) was established through multiple algorithmic approaches. The scale of ICSD goes beyond the confines of any research group could have. This development provides soil for other projects to grow. The Materials Genome Initiative, launched in 2011 in the United States, is a large-scale collaboration between materials scientists (both experimentalists and theorists) and computer scientists to deploy proven computational methodologies to predict, screen, and optimize materials at an unparalleled scale and rate. Materials Project (www.materialsproject.org) is a component of the Materials Genome Initiative that leverages the power of high-throughput computation and practices from the information age to create an open, collaborative, and data-rich ecosystem for accelerated materials design. Based on the large-scale database, data-driven methods could be applied in screening suitable electrode materials such as machine learning. So far, many exciting applications have been developed by data-driven methods, but it is much more than that could be done. We believe that this new approach in the information age could change the paradigm of discovery of new materials.

We hope that through continuous cooperation and efforts and the use of high-resolution characterization tools for in-depth research on the science and engineering of Li metal anodes, there will be more breakthroughs realizing safe and high-performance Li metal batteries in the future. We firmly believe that large-scale practical applications of Li metal batteries are accelerating.

Index

Liquid Electrolyte Chemistry for Lithium Metal Batteries: Design, Mechanisms, Strategies, First Edition. Jianmin Ma.